T0318863

Contagion Phenomena with Applications in Finance

Quantitative Finance Set

coordinated by
Patrick Duvaut and Emmanuelle Jay

Contagion Phenomena with Applications in Finance

Serge Darolles
Christian Gourieroux

First published 2015 in Great Britain and the United States by ISTE Press Ltd and Elsevier Ltd

ISTE Press Ltd
27-37 St George's Road
London SW19 4EU
UK

www.iste.co.uk

Elsevier Ltd
The Boulevard, Langford Lane
Kidlington, Oxford, OX5 1GB
UK

www.elsevier.com

British Library Cataloguing-in-Publication Data
A CIP record for this book is available from the British Library
Library of Congress Cataloging in Publication Data
A catalog record for this book is available from the Library of Congress
ISBN 978-1-78548-035-5

Printed and bound in the UK and US

Contents

Introduction

A large part of the literature on financial contagion and systematic risks has been motivated by the finding that cross-market correlations (respectively, coexceedances) between asset returns increase significantly during crisis periods (see King and Wadhwani [KIN 90] and Bae, Karolyi and Stulz [BAE 03]). Is this increase due to an exogenous shock common to all markets (called interdependence in the literature), or due to certain types of transmission of shocks between markets (called contagion)?

The starting point of this book is that it is not possible to identify the source of these increased co-movements in a purely static framework. Indeed, a static multivariate model for risk variables, such as returns, admit observationally equivalent representations in terms of either a simultaneous equation model, or recursive forms, without the possibility to interpret them in a structural way, for instance, to interpret a recursive form as a causal relationship. This is the reflection problem highlighted by Manski [MAN 93]. To circumvent this difficulty, some authors introduced multivariate models with exogenous switching regimes (see Forbes and Rigobon [FOR 02], Dungey *et al.* [DUN 05]). These models are static in each regime, and correspond to the tranquility regime and the crisis regime, respectively. The basic idea is to allow for a common factor and no contagion in the tranquility regime, whereas the crisis regime can involve contagion and/or extreme values of the common factor. Then, it is possible to identify the main source of increased co-movement by performing standard Chow tests. However, this solution to the identification problem requires identification restrictions, such as the fact that "the common shocks and the idiosyncratic shocks have the same impact during the crisis period as they have during the non-crisis period" [DUN 05, p.11]. These identification restrictions are not necessarily fulfilled empirically and are not testable. Typically, the identification problem will reappear if contagion exists also in the tranquility regime, just at a lesser extent than in the crisis regime. Chapter 1 reviews and discusses all these issues. The discussion shows that we can only expect to identify common factors and contagion within a dynamic framework.

In Chapter 2 we review the standard practices for defining shocks in linear dynamic models with contagion only, called structural vector autoregressive (SVAR) models, and deriving the impulse response functions, that are the dynamic consequences of shocks on the future behavior of the series of interest. Even in a linear dynamic framework, we encounter identification issues when the models are analyzed by second-order approaches. We explain why these identification problems disappear when the errors in the SVAR model are independent and not Gaussian. Moreover, we explain why some noncausal SVAR dynamics are appropriate to capture the speculative bubbles observed in financial markets.

The SVAR model accounts only for contagion, and not for common shocks. We extend the analysis to linear dynamic models with both contagion and common shocks in Chapter 3. To disentangle both effects, a dynamic model requires at least three characteristics: first, the model has to include lagged endogenous variables to represent the propagation mechanism of contagion; second, the specification has to allow for two different sets of factors, namely common factors – also called either systematic factors, global factors [KAN 96], or dynamic frailties [DUF 09] – representing undiversifiable risks, and idiosyncratic factors representing diversifiable risks; finally, these factors have to satisfy exogeneity properties. These exogeneity properties are required for a meaningful definition of shocks on the factors. If a factor is exogenous, a shock on this factor is external to the system and cannot be partially a consequence of contagion. For instance, in a linear dynamic framework, a model disentangling contagion and common factors might be:

$$Y_t = BF_t + CY_{t-1} + u_t, \qquad \text{with } F_t = AF_{t-1} + v_t,$$

where Y_t is the vector of endogenous variables, (u_t) and (v_t) are independent strong white noises, and the components of vector u_t are mutually independent. Vector F_t collects the values at time t of the common factors, and the components of vector u_t are the idiosyncratic factors. The exogeneity of the common factors is implied by the independence assumption between the errors u_t and v_t. Matrix C summarizes the contagion effects. Under standard stability conditions, the previous dynamic model admits a long run equilibrium given by the following static model:

$$Y = BF + CY + u, \quad \text{with} \quad F = AF + v,$$

or equivalently,

$$Y = (Id - C)^{-1}BF + (Id - C)^{-1}u, \quad \text{with} \quad F = (Id - A)^{-1}v.$$

We can now understand why a static model is hopeless for analyzing contagion. Indeed, this last equation corresponds to the long run equilibrium model, provides no information on the trial and error to converge to the equilibrium, and does not allow to identify the contagion matrix C.

Chapter 4 explains how the static and dynamic models of Chapters 1 and 3 can be used for portfolio management, for hedging, or for the analysis of systemic risk in the banking sector.

The linear dynamic models are rather simple, but not able to take into account the nonlinear features existing in risk analysis. These nonlinearities are due to the derivatives traded on the market, to qualitative events such as defaults, prepayments, interventions of a Central Bank, represented by 0-1 indicators. They also have to be introduced to capture risk premia, i.e. the effect of the volatility of a return (a squared return) on the expected return. Chapter 5 extends to nonlinear dynamic models the notion of common factors and contagion. This is illustrated by multivariate models with stochastic volatility.

Chapter 6 introduces a nonlinear dynamic model with both unobservable common factors and contagion for the analysis of the liquidation histories of hedge funds in different management styles.

1

Contagion and Causality in Static Models

A significant part of the academic and applied literature analyzes and measures contagion and causality in a static framework. The aim of this chapter is to review and discuss approaches and to show that such attempts are (almost) hopeless due to identification problems. However, this chapter does not send only a negative message, since it helps to define the notion of shock more precisely: does a shock have to be related to an equation, or to a variable to be systematically aware of identification issues and, last but not least, in noting that such contagion or causality issues can only be considered in a dynamic framework?

1.1. Linear dependence in a static model

The linear dependencies, also called weak dependencies, between variables are the links, which can be detected through the second-order moments, that are the variances and covariances. Equivalently, these are the links, which can be revealed in a Gaussian static framework. For expository purposes, we consider the following Gaussian framework and discuss equivalent specifications of the model, such as its simultaneous equations specification, its recursive forms, but also the specifications underlying exploratory factor analysis and principal component analysis.

1.1.1. *The basic model*

We consider n (random) variables Y_i, $i = 1, \ldots, n$. They can be stacked into a random vector Y of dimension n. This vector is assumed to follow a Gaussian distribution with zero-mean and a variance–covariance matrix Σ:

$$Y \sim N(0, \Sigma). \qquad [1.1]$$

This matrix Σ admits the variances $\sigma_i^2 = V(Y_i)$, $i = 1, \ldots, n$, as diagonal elements, and the covariances $\sigma_{i,j} = cov(Y_i, Y_j)$, $i \neq j$, out of the diagonal.

EXAMPLE 1.1 (Bidimensional case).– For instance, in the bidimensional case $n = 2$, we obtain:

$$\begin{pmatrix} Y_1 \\ Y_2 \end{pmatrix} \sim N\left(0, \begin{pmatrix} \sigma_1^2 & \sigma_{12} \\ \sigma_{12} & \sigma_2^2 \end{pmatrix}\right).$$

1.1.2. *Alternative specifications of the basic model*

There exist alternative ways of writing the static Gaussian model. These alternative specifications make the values of variables Y depend (linearly) on the values of n independently underlying Gaussian variables. Then, these underlying variables are assumed to be the variables on which exogenous shocks will be applied. Moreover, these shocks can be applied separately by the independence property.

1.1.2.1. *Simultaneous equation model*

Gaussian model [1.1] can be rewritten as:

$$Y = CY + \eta, \text{ with } \eta \sim N(0, Id), \qquad [1.2]$$

where the (n, n)-matrix C is assumed to have eigenvalues different from 1, and where the error term η is standard Gaussian. Under specification [1.2], we have the (misleading) impression that a given variable, say Y_i, is affected by the other variables $Y_1, \ldots, Y_n$, plus a noise, and that these effects are summarized in the C matrix. Therefore this C matrix plays a key role for contagion analysis.

However, simultaneous equation system [1.2] is equivalent to the system:

$$Y = (Id - C)^{-1}\eta, \quad [1.3]$$

where Id denotes the identity matrix of size n. We deduce the expression of the variance-covariance matrix Σ as:

$$\Sigma = (Id - C)^{-1}(Id - C')^{-1}. \quad [1.4]$$

Equivalently, $Id - C$ is a "square root" of Σ. But it is well known that a symmetric positive definite matrix has a large number of different square roots. Some square roots are symmetric matrices: there exist 2^n such square roots, whenever Σ has different eigenvalues. There also exist nonsymmetric square roots, for instance square roots, which are lower triangular matrices and are used in the Cholesky decomposition (see e.g. [HIG 01]). Anyway, if we observe independently at several dates the values of the variables Y_t, $t = 1, \ldots, T$, satisfying [1.1], we can expect to accurately approximate matrix Σ by its sample counterpart. But, due to the multiplicity of square roots, we cannot deduce a nonambiguous C; in other words matrix C is not identifiable.

1.1.2.2. *Recursive specification*

Recursive specifications of model [1.1] are frequently considered in practice. For expository purposes, let us discuss the bidimensional case, but the discussion is easily extended to any dimension. For instance, we can write:

$$\begin{cases} Y_1 = u_1, \\ Y_2 = \alpha_{2|1} Y_1 + v_{2|1}, \end{cases} \qquad [1.5]$$

where u_1 and $v_{2|1}$ are independent, $u_1 \sim N(0, \sigma_1^2)$, $v_{2|1} \sim N(0, \sigma_2^2(1-\rho^2))$ and $\alpha_{2|1} = \sigma_{21}/\sigma_1^2$, $\rho^2 = \sigma_{12}^2/(\sigma_1^2 + \sigma_2^2)$.

This recursive specification is a direct consequence of the Bayes formula: the first equation provides the marginal (i.e. unconditional) distribution of Y_1, whereas the second equation provides the conditional distribution of Y_2 given Y_1. $\alpha_{2|1} Y_1$ is the best predictor of Y_2 given as Y_1, which is linear in Y_1 in the Gaussian case; $\sigma_2^2 \sqrt{1-\rho^2}$ is the variance of the associated prediction error. The independence between the two error terms is a consequence of the no correlation between Y_1 and the prediction error $v_{2|1}$ and of the equivalence between no correlation and independence in the Gaussian framework. Specification [1.5] may induce a misleading causal interpretation: variable Y_1 is fixed first (the first equation), then influence Y_2 (the second equation). Such an interpretation is clearly misleading, since it is also possible to get the symmetric recursive representation by changing the orders of indices 1 and 2:

$$\begin{cases} Y_1 = \alpha_{1|2} Y_2 + v_{1|2}, \\ Y_2 = u_2, \end{cases} \text{, say.} \qquad [1.6]$$

A naive view of system [1.6] might lead to detect a causality in the reverse direction from 2 to 1.

Nevertheless, such recursive specifications are still often used in practice for causal interpretations, or for defining the basic shocks when constructing the impulse response functions, that are the changes on Y consequences to some shocks.

For example, Dungey *et al.* [DUN 05] consider the contagion from country 1 to country 2 using the following specification:

$$\begin{cases} y_1 = \lambda_1 w + \delta_1 u_1, \\ y_2 = \lambda_2 w + \delta_2 u_2 + \gamma u_1, \end{cases}$$

where u_1, u_2, w are independent standard normal variables. We have seen in the previous section that this model can also be written as:

$$\begin{cases} y_1 = \lambda_1 w + \alpha_1 v_1 + \gamma v_2, \\ y_2 = \lambda_2 w + \alpha_2 v_2, \end{cases}$$

which gives the impression of a contagion in the reverse direction.

Similarly, Sims [SIM 77, SIM 80] proposed to shock the error terms using the recursive form, say [1.5]. In this scheme, u_1 is interpreted as the shock on variable Y_1, whereas $v_{2|1}$ is interpreted as the shock on the equation defining Y_2. As mentioned earlier, there is no argument to privilegiate this recursive form instead of the second recursive form [1.6]. Nevertheless, this practice leads to important questions:

– how can we define a shock on a variable Y_1?;

– how can we define a shock on an equation?;

– are shocks on variables Y_1, Y_2 (respectively, on equations defining Y_1 and Y_2) independent or linked?;

– if they are linked, how can we represent this link?

1.1.2.3. *Principal component analysis (PCA)*

Any variance–covariance matrix is a symmetric positive definite matrix. This matrix can be diagonalized, that is decomposed as:

$$\Sigma = Q \Lambda Q', \qquad [1.7]$$

where Q is the orthogonal matrix[1] of eigenvectors, Λ the diagonal matrix of associated real eigenvalues (usually written in decreasing order), and all these eigenvalues are strictly positive (see [LAW 71, AND 84]). Thus, we can write:

$$Y = QF, \qquad [1.8]$$

where $F \sim N(0, \Lambda)$. The components of F are called the principal components. These components are Gaussian and uncorrelated. Thus, they are independent and might also be used to define the shocks.

We get a one-to-one relationship between the observations Y and the principal components F. By inverting equation [1.8], we can write the principal components as function of the observations:

$$F = Q^{-1}Y = Q'Y. \qquad [1.9]$$

When Y is a vector of asset returns, the elements of $Q^{-1}Y = Q'Y$ are portfolio returns, corresponding to the portfolio allocations defined by the rows of matrix Q^{-1}, or equivalently by the columns of Q, since $Q^{-1} = Q'$. This provides an interpretation of the components of F as returns of specific portfolios, called (principle components) mimicking portfolios.

To summarize, by simply knowing the unconstrained Σ matrix, it is not possible to identify if the system is causal or simultaneous, and, if it is causal, to identify the causality direction from 1 to 2, or from 2 to 1. It is also not possible to define without ambiguity what has to be shocked either innovations, or principal components, or equations. This is the so-called reflection problem highlighted in [MAN 93].

1 A square matrix Q is orthogonal if it is invertible with $Q^{-1} = Q'$.

1.1.3. *Constrained specifications*

Can we expect an easier analysis when the linear model, that is the matrix Σ, is *a priori* constrained? Below we discuss two constrained specifications.

1.1.3.1. *Exploratory factor analysis*

This type of factor model is the basis of Arbitrage Pricing Theory (APT) (see [ROS 76, CHA 83]). The idea is to capture all the static dependencies through a limited number of factors. The model is written as:

$$\underset{(n,1)}{Y} = \underset{(n,K)}{B}\underset{(K,1)}{F} + \underset{(n,1)}{\epsilon}, \qquad [1.10]$$

where K is the number of factors, B is the matrix of dimension (n, K) of beta coefficients, F is a vector of dimension $(K, 1)$ with F and ϵ independent, $F \sim N(0, Id_K)$, and:

$$\epsilon \sim N\left(0, \begin{pmatrix} \eta_1^2 & 0 & 0 \\ 0 & ... & 0 \\ 0 & 0 & \eta_n^2 \end{pmatrix}\right). \qquad [1.11]$$

Thus, there is no longer correlation between the error terms. This model differs from the model of principal component analysis. First, it contains a number of latent variables, that are the F and ϵ, strictly larger than the dimension of the observations. Second, the links between the observations passes through a limited number K of factors, while PCA will in general exhibit n principal components. More precisely, the model of the explanatory factor analysis implies restrictions on the variance–covariance matrix of the observable variables, when the number K of factors is small with regard to the dimension of Y. Indeed the variance–covariance matrix is given by:

$$\Sigma = BB' + diag(\eta_i^2), \qquad [1.12]$$

where $diag(\eta_i^2)$ denotes the diagonal elements of $diag(\eta_i^2)$, $i = 1, \ldots, n$.

Let us discuss the number of additional restrictions on matrix Σ. In model [1.10]-[1.11], the matrix of beta coefficients and the factors are defined up to an orthogonal matrix Q. Indeed, the model with $B^* = BQ$ and $F^* = Q'F$ provides the same value of the product $B^*F^* = BQQ'F = BF$. It is known that any orthogonal matrix with no eigenvalues equal to 1 can be written as:

$$C = (Id + A)(Id - A)^{-1}, \qquad [1.13]$$

where A is a skew matrix such that $A' = -A$. This is the Cayley's representation of an orthogonal matrix (see [JAC 09]). Thus the number of independent parameters of an orthogonal matrix of dimension K is equal to the number of independent parameters of the skew matrix A, which is $K(K-1)/2$. Model [1.10]-[1.11] contains $n(K+1)-K(K-1)/2$ independent parameters, corresponding to the error variances and the beta coefficients, and taking into account the definition of factor F, up to an orthogonal matrix. There is also one more specification restriction to take into account, since we can always set to zero one of the variance of the errors. Since the dimension of the observable variance–covariance matrix Σ is $n(n+1)/2$, the model is just-identified if:

$$n(K+1) - K(K-1)/2 - 1 = n(n+1)/2,$$

over-identified if:

$$n(K+1) - K(K-1)/2 - 1 < n(n+1)/2,$$

and under-identified, otherwise.

EXAMPLE 1.1 Bidimensional case (continued).– Let us consider the bidimensional case $n = 2$, and a single factor $K = 1$. Thus, the model is just-identified. There is no implied restriction on the variance–covariance matrix of Y and we can write:

$$\begin{cases} Y_1 = \sigma_1 F, \\ Y_2 = \frac{\sigma_{1,2}}{\sigma_1} F + u_2, \end{cases}$$

which is one of the recursive forms.

Model [1.10]-[1.11] is often considered for large dimension n, that is, when n tends to infinity, the number K of factors being fixed. If the components of Y are asset returns, we can consider the equiweighted portfolio, with identical allocations $1/n$ in each asset. Let us denote by e the $n-$dimensional vector with unitary components, $e = (1, \ldots, 1)'$, we have:

$$\frac{1}{n} e'Y = \frac{1}{n} e'BF + \frac{1}{n} e'\epsilon \approx \frac{1}{n} e'BF,$$

for large n, by applying the Law of Large Numbers. Thus, the idiosyncratic terms ϵ play no role, that is, the idiosyncratic risks can be diversified. For large equally weighted portfolios, only the common factors matter.

The argument above can be applied with less diversified portfolios, whenever the allocations $a_1, ..., a_n$, are such that:

$$\lim_{n \to +\infty} \sum_{i=1}^{n} a_i^2 \eta_i^2 = 0.$$

In particular, by writing the asymptotic relationship for a K (linearly independent) diversified portfolio, we see that each component of F is the return of an appropriately selected diversified portfolio, or (factor) mimicking portfolio.

1.1.3.2. *A constrained recursive form*

There exist other constrained specifications defined by means of the recursive form (see [GRA 97]). Let us discuss the case $n = 3$ for expository purpose. With unconstrained Σ, it is always possible to write the model as:

$$\begin{cases} Y_1 = u_1, \\ Y_2 = \alpha_{2|1} Y_1 + v_{2|1}, \\ Y_3 = \alpha_{3|1} Y_1 + \alpha_{3|2} Y_2 + v_{3|1,2}, \end{cases} \qquad [1.14]$$

which is the analog of system [1.5] for $n = 3$. Granger and Swansson [GRA 97] propose to constrain the system and replace it by a "purely" recursive form defined as:

$$\begin{cases} Y_1 = u_1, \\ Y_2 = \alpha_{2|1} Y_1 + v_{2|1}, \\ Y_3 = \alpha_{3|2} Y_2 + v_{3|1,2}, \end{cases} \qquad [1.15]$$

that is, to impose the constraint $\alpha_{3|1} = 0$. This new system gives the impression that we can interpret the link between the variables in terms of the causal scheme: $Y_1 \to Y_2 \to Y_3$. In fact, this constraint is not sufficient to fix the causal scheme without ambiguity, since there exists another causal scheme $Y_3 \to Y_2 \to Y_1$ providing the same volatility–covolatility matrix Σ (see [GRA 97]).

1.2. Nonlinear dependence in a static model

The difficulty of identifying causality and contagion encountered in section 1.1 might be due to the linear (Gaussian) framework. We see below that partial information on causality or systematic factor can be derived in special nonlinear models. This can arise when the observable variables are qualitative, for instance, representing default indicators, or for specific distributional assumptions.

1.2.1. *Recursive specification*

When the observed variables are real with continuous distributions, it is still hopeless to identify the causal directions. This is summarized in lemma 1.1, written below in the bidimensional case for expository framework.

LEMMA 1.1.– Let us consider a pair of variables (Y_1, Y_2) with a bivariate continuous distribution with strictly positive density on $(c_1, \infty) \times (c_2, \infty)$. Then there exist two functions a_1, a_2 such that:

$$Y_1 = a_1(u_1),$$

$$Y_2 = a_2(Y_1, u_2),$$

where u_1 and u_2 are independent standard normal functions, a_1 is strictly increasing with respect to u_1 and function a_2 is strictly increasing with respect to u_2.

PROOF.– See section 1.5.1. □

The nonlinear features appear by the nonlinear transformations of the Gaussian error terms, the nonlinear transformation of Y_1 in the second equation, but also by the possible cross effects between u_2 and Y_1. As in section 1.1, this recursive specification of the static model cannot be interpreted as causal from Y_1 to Y_2. Indeed, the same bivariate system can also be written under the symmetric recursive form:

$$\begin{cases} Y_1 = b_1(Y_2, v_1), \\ Y_2 = b_2(v_2), \end{cases}$$

where v_1, v_2 are independent standard normal variables.

Thus, we cannot identify the causal direction, and it is easily seen that we cannot also distinguish between a causal interpretation and an interpretation with simultaneity.

1.2.2. *Qualitative observations*

The discussion can be different when the observations are qualitative. Let us consider a pair of dichotomous qualitative variables, with values 0 or 1. Their joint distribution is summarized in a contingency table, whose elements $p_{i,j}$, $i, j = 0, 1$, are the joint elementary probabilities. Let us consider the contingency table below, where $p_{0,1} = 0$:

$Y_1 \backslash Y_2$	0	1
0	$p_{0,0}$	0
1	$p_{1,0}$	$p_{1,1}$

When $Y_2 = 1$, then $Y_1 = 1$. But when $Y_1 = 1$, we do not know the value of Y_2.

Due to this asymmetry, Y_2 causes Y_1 when $Y_2 = 1$, but does not cause Y_1 when $Y_2 = 0$. Symmetrically, Y_1 causes Y_2 when $Y_1 = 0$, but does not cause Y_2 when $Y_1 = 1$. Thus a non-rectangular support for the joint distribution can provide partial information on causal directions.

1.2.3. *Factor model*

In a linear framework a factor model is difficult to distinguish from a model without factor in the just identified case $n = 2$, $K = 1$ (see the discussion in example 1.1). This is no longer the case in a nonlinear framework. Let us consider the "linear" factor model:

$$\begin{cases} Y_1 = \beta_1 F + u_1, \\ Y_2 = \beta_2 F + u_2, \end{cases}$$

with independent Gaussian errors, $u_1 \sim N(0, \eta_1^2)$, $u_2 \sim N(0, \eta_2^2)$, but non-Gaussian factor. We assume that this factor has a discrete distribution with two possible values $+1$ and -1, and identical weights: $P(F = 1) = P(F = -1) = 1/2$.

If the observable variables are returns, we have two possible regimes: $F = +1$, for the regime of high returns, and $F = -1$, for the regime of low returns. The pair of variables (Y_1, Y_2) has a continuous bivariate distribution, with joint probability density function (p.d.f.):

$$g(y_1, y_2) = \frac{1}{2}\frac{1}{2\pi\eta_1\eta_2}\exp\left(-\frac{1}{2}\frac{(y_1-\beta_1)^2}{\eta_1^2} - \frac{1}{2}\frac{(y_2-\beta_2)^2}{\eta_2^2}\right)$$
$$+\frac{1}{2}\frac{1}{2\pi\eta_1\eta_2}\exp\left(-\frac{1}{2}\frac{(y_1+\beta_1)^2}{\eta_1^2} - \frac{1}{2}\frac{(y_2+\beta_2)^2}{\eta_2^2}\right).$$

This specification involves four parameters, that is one parameter more than in the basic Gaussian framework of section 1.1. Nevertheless, the existence of a linear factor representation with this specific factor distribution allows identifying all parameters, up to the permutation of indices 1 and 2. The identification is achieved by taking into account the nonlinearities, not only the second-order moments.

To summarize this section, in a nonlinear framework, the identification of either causal directions, or of common factors has to be analyzed separately for each specification.

1.3. Model with exogenous switching regimes

Despite the difficulties in identifying the causality and contagion notions in static models, especially for linear specifications, several authors marginally modified these linear static versions in order to solve the identification issue. For instance, we have seen that the static exploratory factor model generally introduces restrictions on the variance–covariance matrix of the observations and these restrictions can be sufficient to identify the parameters of interest. Moreover, the interpretations in terms of risk diversification and mimicking portfolios support the introduced restrictions.

Other solutions seem less relevant even if they are natural at first sight. The model with exogenous switching regime, largely used in the macro-finance literature, is of this type. It is based on the stylized fact that "the (unconditional) correlations between asset returns increase during the crisis periods" (see [KIN 90]).

Forbes and Rigobon [FOR 02], and Dungey *et al.* [DUN 05] try to answer the following question: is this stylized fact due to interdependence, or due to contagion? Let us see how they defined these notions and look carefully at the identification issue.

1.3.1. *A model with two regimes*

Their idea is to distinguish the quiet (standard) periods from the crisis periods. Let us consider a bidimensional system. During the standard periods, they assume:

$$\begin{cases} Y_1 = \beta_1 F + v_1, \\ Y_2 = \beta_2 F + v_2, \end{cases} \qquad [1.16]$$

where F, v_1, v_2 are independent, $F \sim N(0,1)$, $v_1 \sim N\left(0, \eta_1^2\right)$, $v_2 \sim N\left(0, \eta_2^2\right)$. Thus, they consider an explanatory factor model with a single factor, say. During such periods, the link between the observed variables is measured by the benchmark correlation:

$$corr(Y_1, Y_2) = \frac{\beta_1 \beta_2}{\sqrt{\beta_1^2 + \eta_1^2}\sqrt{\beta_2^2 + \eta_2^2}} = \rho_{12}^*, \; say.$$

Then they assume another model during the crisis periods, with a modified structure of dependence. Two alternative specifications are given in the following.

Alternative model 1

The first alternative model introduces a possible correlation between the error terms. The model is given by:

$$\begin{cases} Y_1 = \beta_1 F + v_1, \\ Y_2 = \beta_2 F + v_2, \end{cases}$$

where $F \sim N(0,1)$ and:

$$\begin{pmatrix} v_1 \\ v_2 \end{pmatrix} \sim N\left(\begin{pmatrix} 0 \\ 0 \end{pmatrix}, \begin{pmatrix} \eta_1^2 & \eta_{12} \\ \eta_{12} & \eta_2^2 \end{pmatrix} \right), \eta_{12} > 0.$$

Thus, the new correlation:

$$\rho_{12} = \frac{\beta_1 \beta_2 + \eta_{12}}{\sqrt{\beta_1^2 + \eta_1^2}\sqrt{\beta_2^2 + \eta_2^2}},$$

takes values between ρ_{12}^* and $\frac{\beta_1 \beta_2 + \eta_1 \eta_2}{\sqrt{\beta_1^2 + \eta_1^2}\sqrt{\beta_2^2 + \eta_2^2}}$.

Alternative model 2

The second alternative model assumes an "increased" common factor. The model becomes:

$$\begin{cases} Y_1 = \beta_1 F + v_1, \\ Y_2 = \beta_2 F + v_2, \end{cases}$$

where $F \sim N(0, \sigma_F^2)$, $\sigma_F^2 > 1$, and $v_j \sim N(0, \eta_j^2)$, $j = 1, 2$, $corr(v_1, v_2) = 0$.

The new correlation:

$$\rho_{12} = \frac{\beta_1 \beta_2 \sigma_F^2}{\sqrt{\beta_1^2 \sigma_F^2 + \eta_1^2}\sqrt{\beta_2^2 \sigma_F^2 + \eta_2^2}} = \frac{\beta_1 \beta_2}{\sqrt{\beta_1^2 + \frac{\eta_1^2}{\sigma_F^2}}\sqrt{\beta_2^2 + \frac{\eta_2^2}{\sigma_F^2}}},$$

takes values between ρ_{12}^* and 1.

1.3.2. *The identification issue*

By considering this exogenous switching regime model with the benchmark specification [1.16] during quiet periods and the modified model, say model 1, during the crisis periods, the identification problem is solved. Indeed, we observe six independent sample variances and covariances (three per regime) for six parameters. The model is just identified and can be estimated and analyzed with standard approaches. For instance, Dungey *et al.* [DUN 05] discuss in detail the procedures to test the hypothesis $H_{01}:\{\eta_{12}=0\}$ for alternative model 1 (or the hypothesis $H_{02}:\{\sigma_F^2 = 1\}$ for alternative model 2).

However, such reasoning is a bit misleading. Indeed, the increased correlation observed during the crisis can be obtained either by an increase of the correlation between v_1 and v_2 (alternative model 1), by an increase of the factor variance (alternative model 2), by an increase of the betas, or by several of these reasons (an alternative model which is not identifiable).

People try to distinguish these different causes of increase and name them:

– shocks affecting the fundamental (i.e. F) [MAS 98];

– specific shocks affecting fundamental series (an increase of $corr(u_1, F)$);

– shocks on betas, or on $V(v)$, called contagion in [FOR 02].

These denominations are rather arbitrary and, last but not least, their existence and magnitude are not identifiable under realistic assumptions.

1.4. Chapter 1 highlights

Causality, contagion and shocks are not identifiable in the static linear model. In order to try to identify some of these notions, constraints have been introduced in the literature, for instance, by means of factor models, constrained recursive specifications, or models with switching regimes. These constraints are difficult to interpret and to justify. Moreover, they usually do not completely solve the identification issues concerning causality and contagion. The only relevant solution to these issues is to consider a dynamic framework.

1.5. Appendices

1.5.1. *Proof of lemma 1.1*

Let us consider the continuous variables Y_1, Y_2 and denote $F_1(y_1) = P[Y_1 < y_1]$, the cumulative distribution function of Y_1, $F_{2|1}(y_2|\, y_1) = P[Y_2 < y_2|\, Y_1 = y_1]$, the cumulative distribution function of Y_2 given $Y_1 = y_1$, and Φ the cumulative distribution function of the standard normal distribution.

i) The variable $u_1 = \Phi^{-1}[F_1(Y_1)]$ is standard normal. We have:

$$\begin{aligned} P[u_1 < u] &= P[\Phi^{-1}(F_1(Y_1)) < u] \\ &= P[Y_1 < F_1^{-1}(\Phi(u))] \\ &= F_1[F_1^{-1}(\Phi(u))] = \Phi(u). \end{aligned} \tag{1.17}$$

ii) Let us now consider the variable $u_2 = \Phi^{-1}[F_{2|1}(Y_2|\, Y_1)]$. We have:

$$\begin{aligned} P[u_2 < u|\, Y_1 = y_1] &= P[\Phi^{-1}[F_{2|1}(Y_2|\, Y_1)] < u|\, Y_1 = y_1] \\ &= P[Y_2 < F_{2|1}^{-1}\,[\Phi(u)|\, Y_1 = y_1]|\, Y_1 = y_1] \\ &= \Phi(u). \end{aligned} \tag{1.18}$$

Since the conditional distribution of u_2 does not depend on y_1, we deduce that u_2 is independent of Y_1. This variable u_2 is also independent of u_1 because of the one-to-one relationship

[1.17] between u_1 and Y_1. We also see that u_2 is standard normal.

The system in lemma 1.1 is then deduced by inverting system [1.17]–[1.18].

2

Contagion in Structural VARMA Models

We first describe the standard practices for defining shocks in structural vector autoregressive (SVAR) models and deriving the impulse response functions, which provide the dynamic consequences of shocks on the future behavior of the series of interest. Then, we explain why identification issues encountered in SVAR models analyzed by second-order approaches disappear, when errors are independent and not Gaussian. Finally, we discuss the SVARMA models with ill-located roots in either the moving-average or the autoregressive dynamics. In particular, we explain why non-causal dynamics are appropriate to capture the speculative bubbles observed on financial markets.

2.1. Shocks in a dynamic model

2.1.1. *The standard practice*

We consider n (random) variables $Y_{i,t}$, $i = 1, \ldots, n$, indexed by t. At date t, they can be stacked into a random vector Y_t of dimension n. The process (Y_t) is assumed to be a stationary solution of the following dynamic simultaneous equation model:

$$\Phi_0 Y_t = \Phi_1 Y_{t-1} + \epsilon_t, \qquad [2.1]$$

where the ϵ_t's are serially uncorrelated, with zero mean: $E\epsilon_t = 0$ and the same variance-covariance matrix: $V\epsilon_t = \Sigma$. Some constraints are usually introduced on Φ_0, Φ_1 and Σ, such as the diagonal elements of Φ_0 are equal to 1.

EXAMPLE 2.1.– As an illustration, let us denote by π_t the inflation rate, g_t the gross domestic product (GDP) growth rate and i_t the nominal interest rate. These three variables may satisfy the following simultaneous equation dynamic model[1]:

$$\begin{cases} \pi_t = 0.9\pi_{t-1} + 0.2g_{t-1} + \epsilon_{1,t}, \\ g_t = 0.3(i_{t-1} - \pi_t) + \epsilon_{2,t}, \\ i_t = 0.9i_{t-1} + 1.5\pi_{t-1} + \epsilon_{3,t}. \end{cases} \qquad [2.2]$$

The equations of such a partly simultaneous system are difficult to interpret. The simultaneity appears through the second equation, which involves both present values of g_t and π_t, and also by means of the possible correlations between the error terms. Clearly, the second equation could also be written as:

$$\pi_t = i_{t-1} - g_t/0.3 + \epsilon_{2,t}/0.3.$$

Therefore, we do not know if the inflation rate is fixed by means of this equation, or by means of the first equation of initial system [2.2], or by means of something else taking into account the dependence between the errors.

1 In practice, the values of the coefficients are unknown and have to be estimated. They are given in the example for expository purposes.

Model [2.1] can be rewritten in a reduced form, more appropriate for statistical inference and impulse analysis:

$$Y_t = \Phi_0^{-1}\Phi_1 Y_{t-1} + \Phi_0^{-1}\epsilon_t. \quad [2.3]$$

EXAMPLE 2.1 (continued).– The corresponding reduced form is:

$$\begin{cases} \pi_t = 0.9\pi_{t-1} + 0.2g_{t-1} + \epsilon_{1,t}, \\ g_t = 0.06g_{t-1} - 0.3i_{t-1} + 0.27\pi_{t-1} + 0.3\epsilon_{1,t} + \epsilon_{2,t}, \\ i_t = 0.9i_{t-1} + 1.35\pi_{t-1} + 0.3g_{t-1} + 1.5\epsilon_{1,t} + \epsilon_{3,t}. \end{cases}$$

The reduced form is usually estimated by ordinary least squares (OLS) equation by equation. This provides consistent estimators of $\Phi_0^{-1}\Phi_1$ and $\Phi_0^{-1}\Sigma(\Phi_0^{-1})'$. However, this approach reveals two identification issues. First, it is not possible to identify the parameters Φ_0, Φ_1 and Σ separately, and then disentangle the simultaneity passing by means of the coefficients in matrix Φ_0, or by means of the correlations between the error terms deduced from matrix Σ. Second, the OLS method systematically provides estimators such that the eigenvalues of the matrix $\Phi_0^{-1}\Phi_1$ have their modulus strictly smaller than 1 (see section 2.5.1).

Until recently, the second issue has been completely neglected and the estimated model usually assumed stable, i.e. with eigenvalues of modulus smaller than 1. The first identification issue is in general solved by triangularization [SIM 80]. Indeed, we do not need identifiable Φ_0, Φ_1 and Σ for determining the impulse response function (IRF), i.e. the dynamic consequences of shocks impacting ϵ_t's on Y_t. For constructing the IRF, we consider an autoregressive specification such as:

$$Y_t = \Phi Y_{t-1} + C\eta_t, \text{ say}, \quad [2.4]$$

where the error terms η_t's are standardized and in particular cross-sectionally uncorrelated: $V\eta_t = Id$. The introduction of the no correlation condition on the error terms is needed to perform shocks on $\eta_{1,t}$, say, with "no impact" on the other components[2] $\eta_{2,t}, ..., \eta_{n,t}$.

Since the matrix Φ is identifiable, the only identification issue concerns the matrix C. Indeed, only $CC' = V_{t-1}(Y_t)$ is identifiable, where $V_{t-1}(Y_t)$ denotes the variance-covariance matrix of Y_t given Y_{t-1} (the so-called volatility-covolatility matrix when the components of Y_t are asset returns). Sims [SIM 80] assumes that C is lower triangular. Then, by Choleski's decomposition, there is a unique C solution to the equation $CC' = V_{t-1}(Y_t)$ (up to the signs of the diagonal elements).

Then the IRF due to a transitory shock on $\eta_{1,t}$ at date T, say, can be derived by computing recursively the effects δY_{T+h} on Y_{T+h} *via* the recursive equation:

$$\delta Y_{T+h} = \hat{\Phi}\delta Y_{T+h-1} + \hat{C}\delta\eta_{T+h-1}, \qquad [2.5]$$

where $\delta Y_{T-1} = 0$, $\delta\eta_T = (\delta\eta_{1,T} \quad 0)'$, $\delta\eta_{T+h} = 0$, $h > 0$, and $\hat{\Phi}$ and $\hat{C}$ are the estimated Φ and C matrices, respectively.

The above approach has a severe drawback: the IRF depends on the ordering of the components of variable Y_t in the triangularization, as well as on the possible preliminary linear transformations performed on the variables. Typically in example 2.1, we will obtain different IRFs, if the triangularization is done with the ranking (π_t, g_t, i_t), or with the ranking (π_t, i_t, g_t), as well as different IRFs if it is done on the nominal variables (g_t, i_t, π_t), or on the real variables $(g_t - \pi_t, i_t - \pi_t, \pi_t)$. How can we select an IRF between these different IRF candidates? There exist several practices such as selecting an IRF with a realistic pattern, or an IRF corresponding to the expectations of the policy maker.

2 We discuss this condition in more detail in section 2.2.1.

Another solution is to introduce additional restrictions with economic interpretations such as short-term causality restrictions on Φ_0, Σ, long-term restrictions on Φ_0, Φ_1, Σ [BLA 89], or sign restrictions on either the autoregressive coefficients or the IRF [UHL 05]. These restrictions are often difficult to justify economically, and do not find consensus between economists. In fact, we will see in section 2.2.2 that, in a better analysis of this SVAR models with non-Gaussian independent errors, these additional restrictions are overidentifying restrictions. Therefore, they can be tested, which is either accepted or rejected.

2.1.2. *Term structure of multipliers*

Even if it is difficult to define the error to be shocked, the effects of such shocks at the different horizons have a same structure. More precisely, let us denote by Δ the effect of such a transitory shock on $C\eta_T$ at date T. Then the effect on Y_{T+h} is computed recursively as:

$$\begin{cases} \delta Y_T &= \Delta, \\ \delta Y_{T+h} &= \hat{\Phi}\delta Y_{T+h-1}, h \geq 1. \end{cases} \quad [2.6]$$

By recursive substitutions, we deduce that:

$$\delta Y_{T+h} = \hat{\Phi}^h \Delta, h \geq 0. \quad [2.7]$$

We note that:

– the effect of the magnitude Δ of the shock on the future values of the series is linear in Δ;

– this effect depends on the horizon h by means of the power of the (estimated) autoregressive matrix.

Thus, for a given series $Y_{i,t}$, say, the effect is:

$$\delta Y_{i,T+h} = (\hat{\Phi}^h)_i \Delta,$$

where $(\hat{\Phi}^h)_i$ denotes the ith row of matrix $\hat{\Phi}^h$. We have:

$$\delta Y_{i,T+h} = \sum_{j=1}^{n} (\hat{\Phi}^h)_{i,j} \Delta_j. \qquad [2.8]$$

The element $(\hat{\Phi}^h)_{i,j}$ gives the sensitivity of $Y_{i,T+h}$ with respect to a shock on the jth component of $C\eta_T$. It is often called a multiplier, and the function $h \to (\hat{\Phi}^h)_{i,j}$ provides the term structure of multipliers.

The pattern of this term structure depends on the eigenvalues of matrix $\hat{\Phi}$. The eigenvalues have a modulus strictly smaller than 1 in our basic specification. They can be real or complex. The positive real eigenvalues imply exponentially decreasing patterns tending to 0 in the long run ($h \to \infty$) (see Figure 2.1).

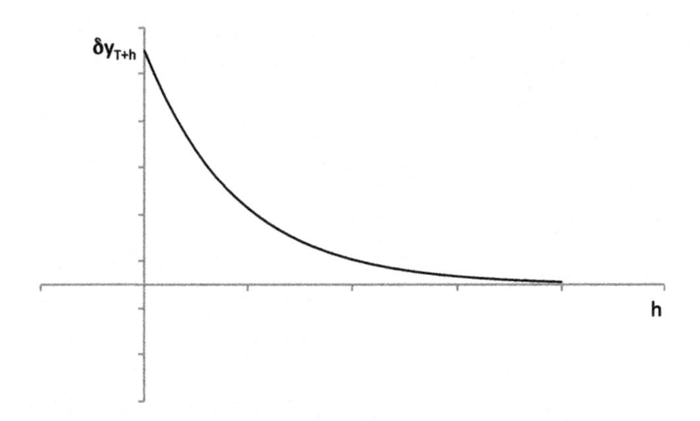

Figure 2.1. *Decreasing term structure pattern*

The negative real eigenvalues imply decreasing periodic patterns of period 2, as displayed in Figure 2.2.

Finally, the complex eigenvalues can be considered in pairs (an eigenvalue and its conjugate) and create amortizing sinusoidal patterns (see Figure 2.3).

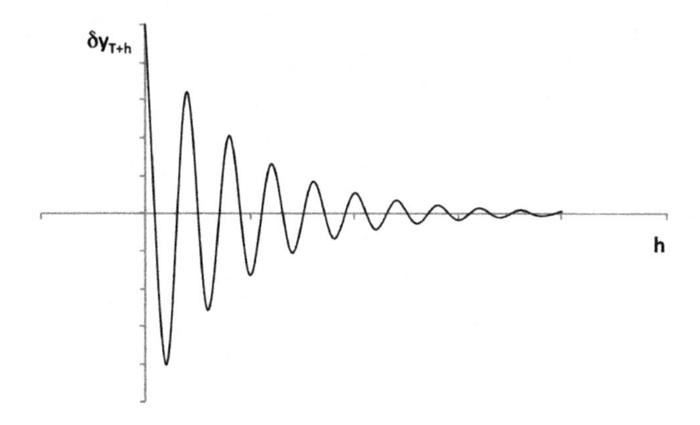

Figure 2.2. *Decreasing periodic term structure pattern (period 2)*

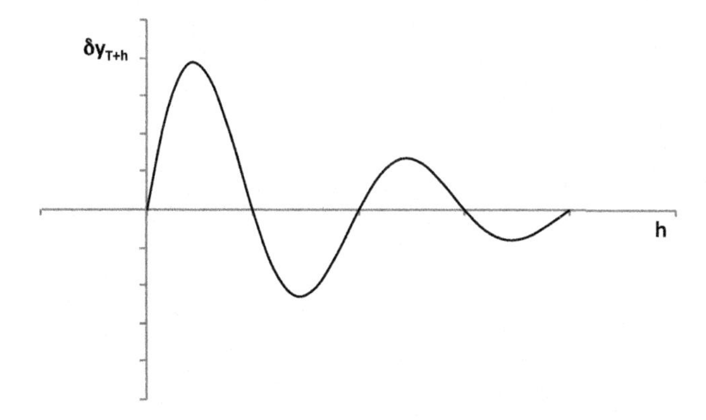

Figure 2.3. *Amortizing sinusoidal term structure pattern*

2.1.3. *Interpretation in terms of contagion and network*

Even if a shock only concerns the error of the first equation, that is if $\Delta_1 \neq 0$, $\Delta_i = 0, i \geq 2$, we generally observe at horizon h an effect on several variables, not only on the first one. To analyze these effects, we have to look at the structure of matrix Φ (or $\hat{\Phi}$), in particular at the zero elements of this matrix and of its different powers. The matrix Φ can be interpreted as summarizing the possible contagions, and the function $h \to \Phi^h$ provides the term structure of contagion. The structure of contagion is usually represented by means of a network between $i = 1, ..., n$ (i.e. the series) with arrows from i to j, if $\phi_{i,j} \neq 0$ (see also [ALL 00]). Let us give some examples of networks for $n = 3$.

EXAMPLE 2.2 (Complete network).– When all the elements of Φ are strictly positive, $\phi_{i,j} > 0, \forall i, j$, we get a network such as in Figure 2.4. The network of horizon h corresponding to Φ^h has the same complete structure.

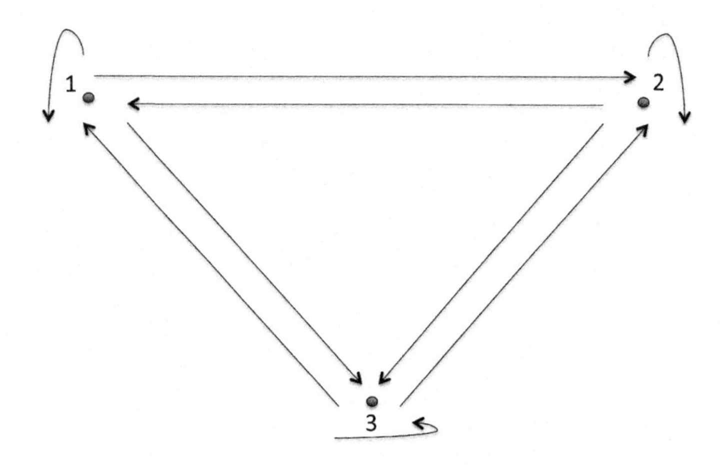

Figure 2.4. *Complete network*

EXAMPLE 2.3 (Recursive network).– Let us now consider a matrix Φ such as:

$$\Phi = \begin{bmatrix} 0 & 0 & \phi_{1,3} \\ \phi_{2,1} & 0 & 0 \\ 0 & \phi_{3,2} & 0 \end{bmatrix}. \quad [2.9]$$

This contagion matrix corresponds to the network displayed in Figure 2.5. At horizon 2, we obtain:

$$\Phi^2 = \begin{bmatrix} 0 & \phi_{1,3}\phi_{3,2} & 0 \\ 0 & 0 & \phi_{2,1}\phi_{1,3} \\ \phi_{3,2}\phi_{2,1} & 0 & 0 \end{bmatrix}.$$

The network is still recursive, but the causal direction between states 2 and 3 is reverted.

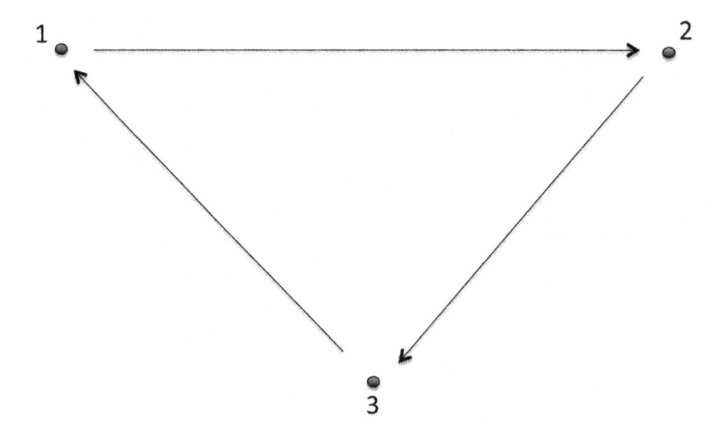

Figure 2.5. *Recursive network*

EXAMPLE 2.4 (Segmented network).– When

$$\Phi = \begin{bmatrix} 0 & \phi_{1,2} & 0 \\ \phi_{2,1} & 0 & 0 \\ 0 & 0 & \phi_{3,3} \end{bmatrix},$$

we get a partition of the network between two segments $(1, 2)$ and (3) with no link between them, as in the scheme given in Figure 2.6.

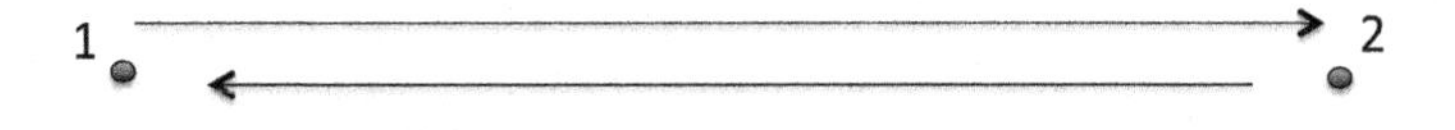

Figure 2.6. *Segmented network*

EXAMPLE 2.5 (Absorbing state).– Finally, some structures can be such that the shock has an effect up to a given finite term on some series, for instance when:

$$\Phi = \begin{bmatrix} 0 \; 0 \; \phi_{1,3} \\ 0 \; 0 \; \phi_{2,3} \\ 0 \; 0 \; \phi_{3,3} \end{bmatrix}.$$

In this example, there is no effect of a transitory shock at date T on series $(Y_{1,t})$, $(Y_{2,t})$ after $T+1$.

2.1.4. *Measure of connectedness*

To illustrate this interpretation in terms of contagion and networks, let us consider an application to the design of econometric measures of connectedness and systemic risk in the finance and insurance sector proposed by Billio *et al.* [BIL 12]. They propose a Granger-causality measure of connectedness to capture the lagged propagation of return spillovers in the financial system, i.e. the network of Granger-causal relations among n financial institutions (banks, insurance companies, brokers or hedge funds).

The degree of Granger causality (DGC) is defined as the fraction of statistically significant Granger-causality relationships among all $n(n-1)$ pairs of n financial institutions:

$$DGC = \frac{1}{n(n-1)} \sum_{i=1}^{n} \sum_{i \neq j} 1_{\phi_{i,j}>0}. \qquad [2.10]$$

The risk of a systemic event is assumed to be high when the DGC exceeds a given threshold. This global connectedness measure can be decomposed as:

$$DGC = \frac{1}{n} \sum_{i=1}^{n} DGC_i(in),$$

where $DGC_i(in) = \frac{1}{n-1} \sum_{j:i \neq j} 1_{\phi_{i,j}>0}$, and also as:

$$DGC = \frac{1}{n} \sum_{j=1}^{n} DGC_j(out),$$

where $DGC_j(out) = \frac{1}{n-1} \sum_{i:i \neq j} 1_{\phi_{i,j}>0}$.

The number of “in” (respectively, “out”) connections, i.e. the number of financial institutions that are significantly Granger-caused by a given institution (respectively, the number of financial institutions that significantly Granger cause a given institution), is in this context a measure of the systemic importance of each single institution. These measures of connectedness can be considered at any term by replacing the elements of Φ by the elements of Φ^h.

2.2. A vector autoregressive moving average (VARMA) model with independent errors

The difficulty in identifying the errors to be shocked and the associated impulse response functions is due to the dynamic analysis essentially based on the second-order moments of the series, which are the variances and the cross and serial correlations. However, to obtain reliable economic interpretations, the shocks have to be made on independent errors, and not only on uncorrelated errors. In this section, we first discuss the difference between the notions of no correlation and independence. Then, we explain that there is neither a static, nor a dynamic identification problem if the errors are cross-sectionally and serially independent, and not Gaussian.

2.2.1. *No correlation versus independence*

It is well known that two independent variables η_1, η_2, say, are necessarily uncorrelated, and that the notions of no correlation and independence are equivalent if the joint vector $(\eta_1, \eta_2)'$ is Gaussian. What can be said when one of these variables is not Gaussian? An analysis based only on the correlation can be very misleading as seen in the example below.

Let us consider two variables η_1, η_2 such that $\eta_2 = \eta_1^2$, and η_1 is standard normal: $\eta_1 \sim N(0,1)$. We have:

$$Cov(\eta_1, \eta_2) = E(\eta_1^3) - E(\eta_1)E(\eta_1^2) = 0.$$

Therefore, these variables are uncorrelated. This result is easily understood if we consider the support of the joint distribution of $(\eta_1, \eta_2)'$ (see Figure 2.7). Due to the symmetry of the standard normal distribution and the parabolic form of the support, the regression line is parallel to the $x-$axis, that is, the correlation coefficient is equal to zero.

This example is rather extreme, since these uncorrelated variables are in a deterministic relationship. In particular, we cannot shock η_1 without shocking $\eta_2 = \eta_1^2$, even if the variables are uncorrelated.

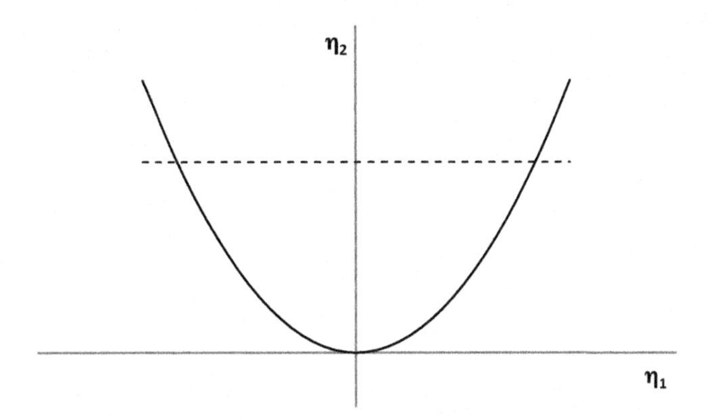

Figure 2.7. *Support of the joint distribution and regression line. The figure displays with the support of the joint distribution (continuous line) and the regression line (dotted line)*

2.2.2. *The identification result*

Let us consider a stationary process with a two-sided average representation:

$$Y_t = \sum_{j=-\infty}^{+\infty} A_j \eta_{t-j}, \qquad [2.11]$$

where the errors (η_t) are independent, identically distributed and the component $\eta_{i,t}, i = 1, ..., n$ is cross-sectionally independent, with $V(\eta_{i,t}) = 1, i = 1, ..., n$. Such a specification includes the pure causal process, when only the current and past values of the noise matter:

$$Y_t = \sum_{j=0}^{+\infty} A_j \eta_{t-j}, \qquad [2.12]$$

as well as the pure non-causal process:

$$Y_t = \sum_{j=-\infty}^{0} A_j \eta_{t-j}.$$

However, as seen in the next sections, processes with both non-degenerated causal and non-causal components may exist in practice. The following result solves the identification issue for such two-sided moving average processes (see [CHA 06])[3].

PROPOSITION 2.1 (Identification of a moving average).– The moving-average matrix coefficients $A_j, j = -\infty, ..., +\infty$ are identifiable (up to a change of sign on the components of η_t and to a permutation of the components), if at most one component $\eta_{i,t}$ is Gaussian.

This result can in particular be applied to the static model of Chapter 1 and to the $VAR(1)$ model of section 2.1. Let us consider the static framework. The reduced form [1.3] is:

$$Y_t = (Id - C)^{-1}\eta_t = A\eta_t, \quad \text{say.}$$

The identification result says that the matrix A is identifiable if the components $\eta_{i,t}, i = 1, ..., n$ are independent, with unitary variance, and non-Gaussian. Generally speaking, this matrix can be estimated consistently. As a result, there is no need to look for an *ad-hoc* triangularization *à la* Sims. It is even possible that the true matrix A estimated from the data is not triangular at all, even after a permutation of the components of η.

Therefore, it is important to check if the series of interest is Gaussian or not, to see if we can expect the identification property above to be applied. As an example, we provide in Figure 2.8 the historical distributions of the inflation rate, the change in GDP and a short-term rate in US. The distribution of inflation rate and change in GDP are skewed. The distribution of short-term interest rate is bimodal, with a mode close to zero, corresponding to the period with low interest rates. These distributions are clearly not Gaussian.

3 A similar identification result has been derived when the errors have fat tails as usual in economy and finance in [GOU 15d].

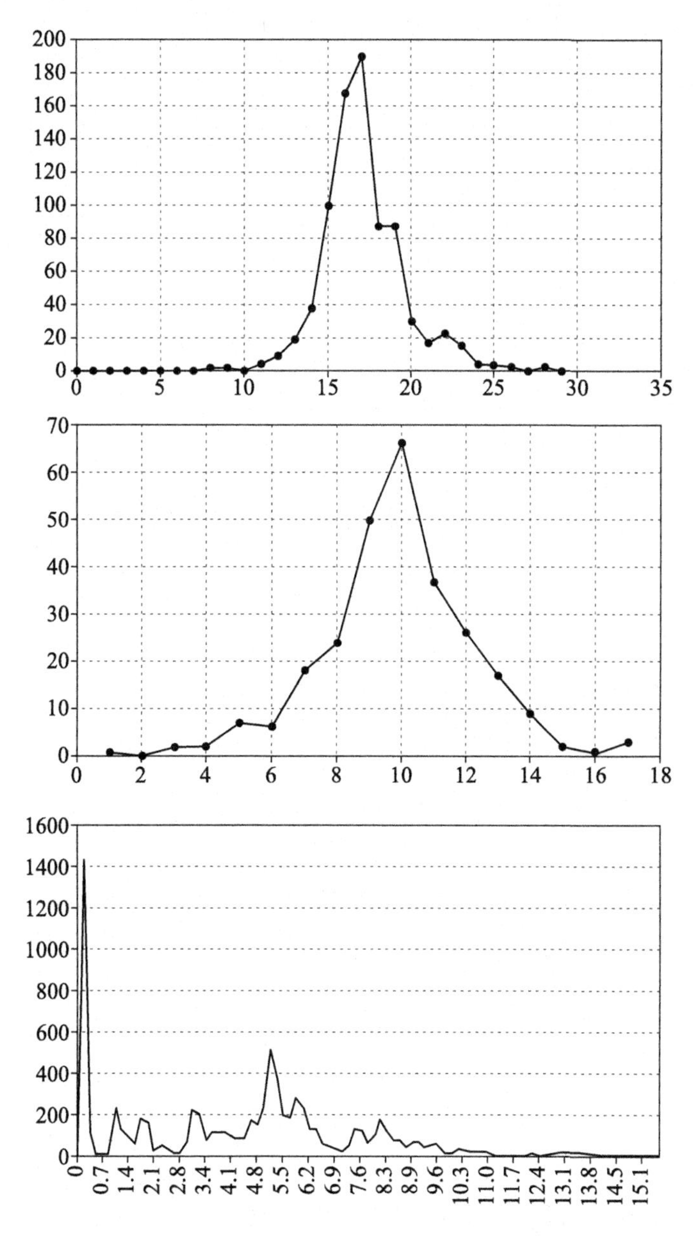

Figure 2.8. *Historical distributions. This figure displays the historical distributions of the inflation rate (top), the change in GDP (middle) and a short-term rate in the US (bottom)*

2.3. Non-fundamentalness

2.3.1. *Ill-located roots*

When the errors are Gaussian, the two-sided moving average process defined in [2.11] also admits a one-sided moving average representation [2.12]. The identification property shows that this cannot arise in the non-Gaussian case.

How should we interpret the previous results when in practice the infinite moving average representation is approximated by a VARMA representation? For expository purposes, let us consider a VARMA(1,1) model with autoregressive and moving average orders both equal to 1:

$$Y_t = \Phi Y_{t-1} + \Theta_0 \eta_t + \Theta_1 \eta_{t-1}, \qquad [2.13]$$

where the η_t's are serially independent, identically distributed and also cross-sectionally independent, with $E\eta_t = 0, V\eta_t = Id$. This model can be alternatively written as:

$$(Id - \Phi L)Y_t = (\Theta_0 + \Theta_1 L)\eta_t, \qquad [2.14]$$

where L denotes the lag operator which changes a process (Y_t) into the process $(LY_t) = (Y_{t-1})$. When the roots of the determinant $det(Id - \Phi z)$ are well located, that is when their modulus is strictly larger than 1, the process (Y_t) admits a causal moving average representation. When the roots of the determinant $det(\Theta_0 + \Theta_1 z)$ are well located, the errors η_t are equal to the innovations of the process: $\eta_t = Y_t - E(Y_t - 1|\underline{Y_{t-1}})$, with $\underline{Y_{t-1}} = \{Y_{t-1}, Y_{t-2}, ...\}$.

The VARMA representation is fundamental if the roots are well located for both the autoregressive and the moving average polynomials. Thus, in a VARMA model, there may exist two problems of non-fundamentalness, that are ill-located roots. They can concern either the moving average polynomial (we say that the VARMA process is not invertible) and/or the autoregressive polynomial (we say that the process is mixed causal/non-causal).

2.3.2. *Non-invertible process*

Let us focus on a VARMA process [2.13] with well-located roots on the autoregressive polynomial and possibly ill-located roots on the moving average polynomial. Typically, for $n = 1$, equation [2.13] becomes:

$$y_t = \phi y_{t-1} + c\eta_t + c\theta\eta_{t-1}, |\phi| < 1, |\theta| > 1. \qquad [2.15]$$

Different economic arguments have been given in the literature for such ill-located roots of the moving average polynomial (see [ALE 11] and [GOU 15b] for a general discussion). This can be justified by a lag between the occurrence of a shock and its maximum effect on y_t. The shock can be a productivity shock [LIP 93], or a change in fiscal policy [LEE 13]. Such non-invertible processes can also arise in rational expectation models. For example, Hansen and Sargent [HAN 91] consider the following model:

$$y_t = E_t\left(\sum_{h=0}^{+\infty} \beta^h w_{t+h}\right), \qquad [2.16]$$

with $w_t = \epsilon_t - \theta\epsilon_{t-1}$, $0 < \beta < 1$ and $|\theta| < 1$. Thus, the conditions of stability are satisfied on both β and θ. If the available information[4] is $I_t = (\epsilon_t, y_t, \epsilon_{t-1}, y_{t-1}, \ldots)$, the solution of the rational expectation model [2.16] is:

$$y_t = (1 - \beta\theta)\epsilon_t - \theta\epsilon_{t-1}, \qquad [2.17]$$

4 This is the information of the economic agent. Ex-post, it will differ from the information of the econometrician equal to $(y_t, y_{t-1}, \ldots)$.

where the root of the moving average (MA) polynomial is $(1 - \beta\theta)/\theta = 1/\theta - \beta$. This root can be larger or smaller than 1 according to the parameter values (see Figure 2.9).

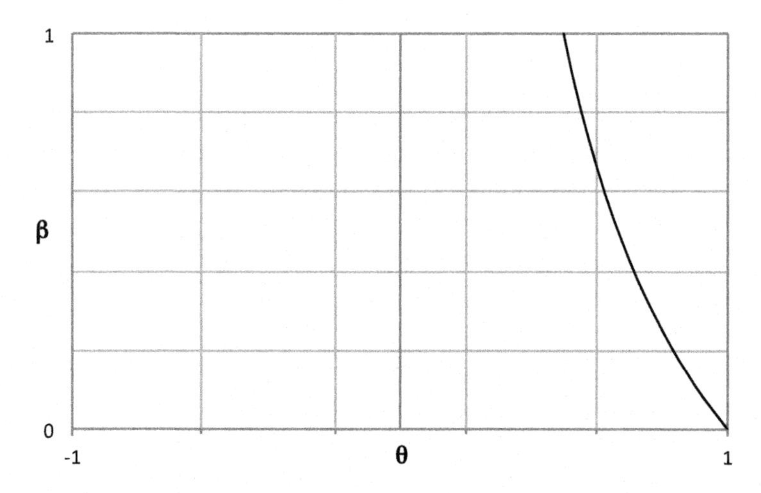

Figure 2.9. *The invertibility region. This figure displays with a continuous line the couples (θ, β) satisfying the unitary condition: $1/\theta - \beta = 1$. The invertibility region is located below this continuous line*

The computation of the impulse response functions for the VARMA specification [2.13]:

$$Y_t = \Phi Y_{t-1} + C_0 \eta_t + C_1 \eta_{t-1},$$

can be done in a rather standard way from data on $Y_1, ..., Y_T$, and estimated parameters $\hat{\eta_T}$, $\hat{\Phi}$, $\hat{C}_0$, $\hat{C}_1$, $\hat{f}_1$, ..., $\hat{f}_n$, where $\hat{f}_i$ is the estimated p.d.f. of $\eta_{i,t}$.

Note, however, that:

1) The errors to be shocked are some components of η, since these errors may have an economic interpretation. The shocks are not made on the innovations of the process which differ from the η's in the non-invertible case.

2) It is equivalent to shock $\eta_{i,T}$, say, or to shock its distribution. For instance, a change from $\eta_{i,T} \to \eta_{i,T} + \delta_i$ is equivalent to a drift of δ_i on the distribution f_i.

3) Since the distribution can be estimated, we can compare the simulated future paths with and without shocks. By replicating these paths, we cannot only compute the average effect of the shock, but can also compute an interval at 90%, say, for this effect, by considering the associated quantiles at 5% and 95% levels, say, as the lower and upper bounds of this interval.

Without shock, the first step is to simulate $\eta^s_{T+1}, \dots, \eta^s_{T+h-1}$ from $\hat{f}_1, \dots, \hat{f}_n$, and then deduce simulated paths $Y^s_{T+1}, \dots, Y^s_{T+h}, \dots$ by applying recursively:

$$Y^s_{T+h} = \hat{\Phi} Y^s_{T+h-1} + \hat{C}_0 \hat{\eta}^s_{T+h} + \hat{C}_1 \hat{\eta}^s_{T+h-1}, \quad h \geq 1. \qquad [2.18]$$

With shocks, the same principle is followed, but with drawings in shocked distributions $\hat{f}^c_1, \dots, \hat{f}^c_n$, for instance. There exist a lot of more or less relevant alternatives to define the shocked distributions. For instance, we may:

1) introduce a drift on distribution f_1, which is the deterministic shock on η_1;

2) change the variance of distribution f_1 to analyze the consequence of an increase in an idiosyncratic factor;

3) introduce a correlation between η_1 and η_2 along the idea of Forbes, Rigobon described in Chapter 1.

2.3.3. *Mixed causal/non-causal process*

Let us now consider the problem of ill-located roots in the autoregressive dynamics. For expository purposes, let us focus on a stationary autoregressive model of order 1:

$$Y_t = \Phi Y_{t-1} + C_0 \eta_t, \qquad [2.19]$$

where the eigenvalues of Φ have a modulus different than 1, but this modulus can be smaller than 1 or larger than 1.

As already mentioned, the standard OLS approach for estimating matrix Φ can lead to inconsistent results and has to be avoided. Nevertheless, there exist consistent methods such as the maximum likelihood method when the distribution of the errors is parametrically specified (see, e.g. [BRE 91, DAV 12] and [LAN 13]) or non-parametric approaches based on estimated covariances (see, e.g. [GOU 15a] and [GOU 15b]). When these consistent methods are applied to economic and financial series, we often obtain estimated $\hat{\Phi}$ matrices with eigenvalues whose modulus is larger than 1. How can we explain the presence of non-causal linear features in these series?

To understand this stylized fact, it is important to consider an alternative interpretation of a two-sided moving average process. Let us consider the one-dimensional case $n = 1$. The process [2.11] can be written as:

$$y_t = \sum_{\tau=-\infty}^{+\infty} \eta_\tau a_{t-\tau}. \qquad [2.20]$$

Thus, the path of process (Y_t) is a combination of baseline paths $Z_\tau(t) = a_{t-\tau}$, $\tau = -\infty, \ldots, +\infty$ with stochastic coefficients η_τ.

As an illustration, we provide in Figure 2.10 the baseline paths for a causal autoregressive (AR) process: $a_h = \rho^h$, for $h \geq 0$ (left panel), and a non-causal process: $a_{-h} = \rho^h$, for $h \geq 0$ (right panel). For a causal AR process, the observed path is a superposition of the basic paths of the left panel, and such a basic pattern can be revealed if there is one η_τ drawn in the tail of the distribution. Thus, we can expect to observe jumps followed by an exponential decrease.

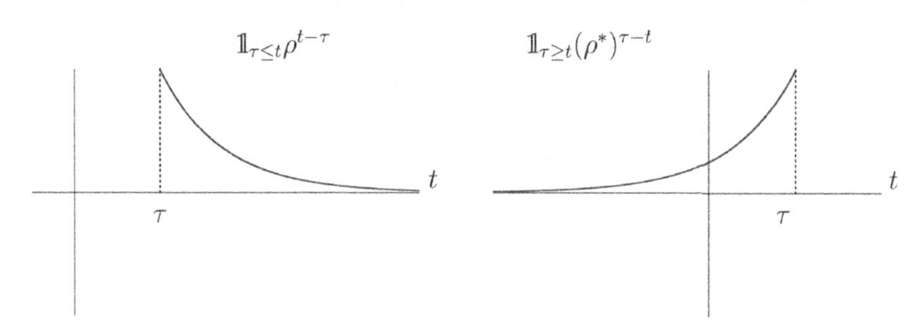

Figure 2.10. *Baseline paths for causal and non-causal AR processes*

Symmetrically, for a non-causal process, a drawing in the tail of the distribution of η can reveal the basic pattern described in the right panel of Figure 2.10, which is an exponential increase followed by a crash. This is typically the pattern of a so-called speculative bubble.

Thus, the observed presence of non-causal dynamic features is likely a consequence of the frequent speculative bubbles present in financial and economic time series. In their pioneering paper, Gourieroux and Zakoian [GOU 15c] have studied in detail a non-causal autoregressive process with Cauchy errors:

$$y_t = \phi y_{t-1} + \eta_t,$$

where $\phi > 1$ and η_t follows a Cauchy distribution.

Since η_t has no expectation, the expectation of y_t does not exist, as well as the backward expectation $E(y_t|y_{t+1})$ and the correlation $Corr(y_t, y_{t+1})$. Nevertheless, they prove that the forward expectation $E(y_t|y_{t-1})$ exists and is such that:

$$E(y_t|y_{t-1}) = y_{t-1}. \quad [2.21]$$

Therefore, this process without first unconditional moment is a (generalized) martingale.

As a result, this type of process can appear as feasible solutions in rational expectation equilibrium models (see section 2.5.2), and the speculative bubbles are a consequence of self-fulfilling prophecies.

The example of the non-causal Cauchy process shows that we have to distinguish two types of "unit root" models:

1) a random walk model, where the process satisfies:

$$y_t = y_{t-1} + \epsilon_t, \text{ say},$$

with ϵ_t independent of $\underline{y_{t-1}}$. This is a non-stationary process, able to capture "permanent" explosion phenomena;

2) a non-causal Cauchy process:

$$y_t = \varphi y_{t-1} + \eta_t, \varphi > 1,$$

which is stationary. The error η_t is linked to $\underline{y_{t-1}}$. This type of process is able to capture "transitory" explosions, followed by a crash.

2.4. Chapter 2 highlights

In a linear framework, pure contagion models are generally represented by VARMA processes. To get reliable definition of shocks and to analyze their effects on the variables of interest, it is necessary to assume that the errors are both serially and cross-sectionally independent. Indeed, approaches based on first-order and second-order moments suffer from both static and dynamic identification issues. Finally, we discussed the problems of non-fundamentalness in VARMA processes, and in particular the importance of non-causal autoregressive processes to represent speculative bubbles.

2.5. Appendices

2.5.1. *A property of the causal OLS estimator in VAR(1) model*

2.5.1.1. *The property*

Let us consider a stationary VAR(1) model:

$$Y_t = \Phi Y_{t-1} + \epsilon_t,$$

where (ϵ_t) is a sequence of i.i.d. random vectors, with second-order moments, and the autoregressive matrix Φ does not necessarily have all its eigenvalues with a modulus strictly smaller than 1.

In practice, the autoregressive matrix Φ is generally estimated by regressing the components of Y_t on the components of Y_{t-1}. This practice can be inappropriate and highly misleading. Indeed, it is based on causal regressions, whereas the process can have non-causal components. The limiting value of the OLS estimator is:

$$\Phi_\infty = Cov(Y_t, Y_{t-1})V(Y_{t-1})^{-1}.$$

We have the following property.

PROPOSITION 2.2.– The eigenvalues of Φ_∞ have a modulus strictly smaller than 1.

Thus, the standard OLS method leads to an inconsistent estimator of the true autoregressive matrix, if the process includes non-causal components.

2.5.1.2. *Proof of proposition 2.2*

Let us consider an eigenvector u of the matrix Φ_∞ associated with the eigenvalue λ. We have:

$$\Phi_\infty u = \lambda u.$$

By taking the conjugate and transpose, we obtain:

$$\bar{u}'\Phi'_{\infty} = \bar{\lambda}\bar{u}'.$$

We deduce:

$$\bar{u}'\Phi'_{\infty}\Phi_{\infty}u = |\lambda|^2 \bar{u}'u.$$

In particular, $|\lambda|^2$ is smaller than the largest eigenvalue of the symmetric matrix $\Phi'_{\infty}\Phi_{\infty}$. Without loss of generality, we can assume $V(Y_t) = Id$ (or equivalently the data are historically prewhitened). We have:

$$\Phi'_{\infty}\Phi_{\infty} = Cov(Y_{t-1}, Y_t)Cov(Y_t, Y_{t-1}).$$

Let us now consider the limiting value of the variance-covariance matrix of the (causal) OLS residuals. This matrix is:

$$\begin{aligned} V(\hat{\epsilon}_t) &= V(Y_t) - Cov(Y_t, Y_{t-1})V(Y_{t-1})^{-1}Cov(Y_{t-1}, Y_t), \\ &= Id - Cov(Y_t, Y_{t-1})Cov(Y_{t-1}, Y_t) = Id - \Phi_{\infty}\Phi'_{\infty}. \end{aligned} \quad [2.22]$$

The eigenvalues of a variance-covariance matrix are positive; therefore, all eigenvalues of $\Phi_{\infty}\Phi'_{\infty}$ are strictly smaller than 1. The proof is completed by noting that $\Phi_{\infty}\Phi'_{\infty}$ and $\Phi'_{\infty}\Phi_{\infty}$ have the same eigenvalues.

2.5.2. *Non-causal solutions of a rational expectation equilibrium model*

Mixed processes appear when solving rational expectation (RE) equilibrium models. Let us consider the RE model introduced by Diba and Grossman [DIB 88]:

$$y_t = aE_t(y_{t+1}) + \epsilon_t, a > 0. \quad [2.23]$$

If $a < 1$, there exists a unique stationary solution $y_t^0 = \epsilon_t$. If $a > 1$, there exists an infinite number of stationary solutions, which are combinations of the forward solution $y_t^0 = \epsilon_t$, and the perfect foresight solution [GOU 82]:

$$y_t^1 = \frac{L}{L - a}\epsilon_t. \qquad [2.24]$$

This result is modified if we allow for stationary solutions without first-order moment. The RE equilibrium model is obtained by matching demand and supply and involves two economic shocks: ϵ_t, w_t say. It can be shown that, even if $a < 1$, there exists an infinite number of stationary solutions. For instance, if ϵ_t and w_t are independent, we get the following solutions:

$$y_t = \epsilon_t + y_t^*, \qquad [2.25]$$

with $y_t^* = \rho y_{t+1}^* + \epsilon_t^*$, where ϵ_t^* and ϵ_t are independent white noises, ϵ_t^* is an appropriate nonlinear function of past, current and future values of w_t, and $|\rho| < 1$. In this case, the VAR model:

$$\begin{cases} y_t = \epsilon_t + y_t^*, \\ z_t = \epsilon_t, \end{cases}$$

where $y_t^* = \rho y_{t+1}^* + \epsilon_t^*$, $|\rho| < 1$, cannot be directly used for deriving impulse response functions. Indeed, ϵ_t^* is a function of the future values $Y_t, Y_{t+1}, \ldots$ and has no economic interpretation. The errors with structural interpretations are ϵ_t and w_t. The dynamic system can be rewritten in terms of the structural shocks as:

$$\begin{cases} y_t = \epsilon_t + y_t^*, \\ y_t^* = g(y_{t-1}^*, w_t; s, \rho), \\ z_t = \epsilon_t, \end{cases}$$

with a nonlinear function g, which depends on the tail magnitude of the error distribution. This causal nonlinear representation has to be used for deriving the impulse response functions (see [GOU 15b]). Thus, the mixed causal/non-causal linear dynamic model implies a nonlinear impulse response function.

3

Common Frailty versus Contagion in Linear Dynamic Models

From Chapter 1 we cannot expect to disentangle the effects of common factors and the contagion effects in a static framework without introducing identification restrictions difficult to justify in practice. The aim of this chapter is to show that this identification issue can be solved in a dynamic framework, even within a linear specification.

We describe and discuss the specification in section 3.1. We derive the stationarity condition for the process of observations, discuss the curse of dimensionality problem and its practical implications in terms of autoregressive lags. The common factors play a key role in the specification. They can be assumed to be either observable or latent. We compare in section 3.2 the two approaches. Whereas a specification with latent factors is more complicated to estimate and analyze, it is more appropriate for the prediction and management of future risks. We explain in section 3.3 how to shock the specific and systematic factors and to derive the associated impulse response functions. These impulse response functions can be used to perform stress tests, to derive ratings for systemic risk and to detect the systemically important financial institutions (SIFIs) (see [FIN 13]). Very often the applied studies are based on too constrained models, which disregard either common factors, or contagion. Such constrained analyses are generally misspecified and imply both biased measures of contagion and biased measures of systematic risks. We discuss these biases in section 3.4. The applied literature is reviewed in section 3.5. Technical definitions and proofs are gathered in the appendices (section 3.7).

3.1. Linear dynamic model with common factor and contagion

3.1.1. *The basic specification*

To disentangle the effects of common shocks and the contagion effects, a dynamic model requires at least three characteristics. First, the model has to include lagged variables to identify the propagation mechanism. Second, the specification has to allow for different factors, that are a set of common factors representing undiversifiable risks and a set of idiosyncratic factors representing diversifiable risks. Finally, the factors have to satisfy exogeneity properties. These exogeneity properties are required for a meaningful definition of shocks on the factors. If a factor is exogenous, a shock on this factor is external to the system and cannot be partially a consequence of contagion (see section 3.7.1 for a formal definition of causality and exogeneity in a dynamic model).

For instance, in a linear dynamic framework, such a model might be:

$$Y_t = BF_t + CY_{t-1} + u_t, \text{ with } F_t = AF_{t-1} + v_t, \qquad [3.1]$$

where Y_t is the $n-$dimensional vector of endogenous variables, u_t, v_t are independent strong white noises of dimensions n and K, respectively, and the components of u_t are independent such that:

$$V(u_t) = \sigma^2 Id_n,\ V(v_t) = \Omega,\ Cov(u_t, v_t) = 0. \qquad [3.2]$$

F_t denotes the values at time t of the K common factors and the components of u_t are the idiosyncratic factors. The exogeneity of the common factors is implied by the independence assumption between the errors u and v. Equation [3.1] shows that the endogenous variables Y_t are dynamically dependent through the effects of the common factors, measured by the (n, K) matrix B of beta or sensitivity coefficients, and through the effects of their lagged values, measured by the (n, n) matrix C that summarizes the contagion effects. For expository purposes, we have not introduced an intercept in equation [3.1]. Indeed, whenever the process (Y_t) is stationary, this intercept can be set to zero by considering the demeaned return (or demeaned change in price) $Y_t - \overline{Y}$.

Dynamic model [3.1] can also be written by stacking vectors Y_t and F_t. After substitution of the expression of F_t in the equation defining Y_t, we get:

$$Y_t = BAF_{t-1} + CY_{t-1} + u_t + Bv_t, \text{ with } F_t = AF_{t-1} + v_t,$$

or equivalently:

$$\begin{pmatrix} Y_t \\ F_t \end{pmatrix} = \begin{pmatrix} C & BA \\ 0 & A \end{pmatrix} \begin{pmatrix} Y_{t-1} \\ F_{t-1} \end{pmatrix} + \begin{pmatrix} u_t + Bv_t \\ v_t \end{pmatrix}. \qquad [3.3]$$

We get a Vector AutoRegressive (VAR) representation with observable and unobservable components, Y and F, respectively. This type of dynamic model is known under the acronym of FAVAR for factor augmented vector autoregressive model (when F, or a proxy of F, is observed) (see [BER 05]). Moreover, the VAR representation [3.3] includes constraints on the autoregressive matrix and on the variance–covariance matrix of the error term. These constraints, such as the null South-West block of the autoregressive matrix, result from the exogeneity assumption and the need to get independent idiosyncratic factors. This FAVAR representation is useful to discuss the existence of stationary solutions for both Y and F. Such a stationary solution exists, if system [3.3] is stable, that is, if the autoregressive matrix in complete system [3.3] admits all its eigenvalues with a modulus strictly smaller than 1. Throughout this section[1], we assume that the joint process is causal. It is easily checked that these eigenvalues are obtained by putting together the eigenvalues of matrices C and A. Indeed the eigenvalues of the autoregressive matrix in [3.3] are the solutions λ of:

$$\det\left[\begin{pmatrix} C & BA \\ 0 & A \end{pmatrix} - \lambda Id_{n+K}\right] = 0$$

$$\Leftrightarrow \det\begin{pmatrix} C - \lambda Id_n & BA \\ 0 & A - \lambda Id_K \end{pmatrix} = 0$$

$$\Leftrightarrow \det\left(C - \lambda Id_n\right)\det\left(A - \lambda Id_K\right) = 0,$$

by using the expression of the determinant of a block decomposed matrix:

$$\det\begin{pmatrix} \Psi_{11} & \Psi_{12} \\ \Psi_{21} & \Psi_{22} \end{pmatrix} = \det\left(\Psi_{11}\right)\det\left(\Psi_{22} - \Psi_{21}\Psi_{11}^{-1}\Psi_{12}\right),$$

applied in the special case $\Psi_{21} = 0$. Thus the system is causal if both the eigenvalues of A and C have modulus strictly smaller than one.

Under these stability conditions and if the future values of the errors are crystallized to their current values $u_t = u$, $v_t = v$,

1 See section 2.3.3 for a discussion of mixed causal/noncausal VAR.

say, the dynamic model above admits a long run equilibrium at date t given by the following static model:

$$Y = BF + CY + u, \text{ with } F = AF + v,$$

or equivalently,

$$Y = (Id - C)^{-1}BF + (Id - C)^{-1}u, \text{ with } F = (Id - A)^{-1}v. \quad [3.4]$$

We can now understand why a static model is hopeless for analyzing contagion. Indeed, the static model corresponds to the long-run equilibrium, provides no information on the Walrasian tâtonnement toward the equilibrium and the contagion matrix C cannot be identified from the static version.

3.1.2. *Extensions*

Let us introduce the lag operator L, which transforms a series (Y_t) into the series $(LY_t) = (Y_{t-1})$. Thus, this operator applies a unitary backward drift on time. Model [3.1] can be extended by introducing more lags in the specification. The extended model is:

$$Y_t = B(L)F_t + C(L)Y_{t-1} + u_t, \text{ where } F_t = A(L)F_{t-1} + v_t, \quad [3.5]$$

where the lag-polynomials $A(L)$, $B(L)$ and $C(L)$ have degrees p_A, p_B and p_C, respectively. Such an extended version allows for a superposition of factor (respectively contagion) effects generated at several previous dates. However, this extension can only be used with small autoregressive orders due to the curse of dimensionality, especially for large cross-sectional dimension n. Indeed the number of observed Y variables is nT, for a number of unknown autoregressive parameters equal to $p_B nK + p_C n^2 + p_A K^2$. This difficulty is illustrated in Table 3.1.

Factors and lags\Assets	10	30	100
$K = 3$, $p_A = p_B = p_C = 1$	139	999	10,309
$K = 5$, $p_A = p_B = p_C = 1$	175	1,075	10,525
$K = 5$, $p_A = p_B = p_C = 2$	350	2,150	21,050

Table 3.1. *Number of assets versus number of parameters*

We may encounter a number of parameters of the same order as the number of observations, or even larger. Typically, for 100 observed asset returns, $K = 5$, $p_A = p_B = p_C = 2$, the number of unknown parameters, equal to $21,050$ (which does not account for the parameters in the variance–covariance matrix of the error terms u_t and v_t), requires several hundreds of observation dates. Because of the curse of dimensionality, it seems difficult to significantly increase the number of lags, especially for the contagion effect, that is, for p_C.

Finally, we may include a vector intercept in model [3.1], or in its extension [3.5]. But, instead of considering a model with intercept applied to the raw data series, it is often preferred in practice to demean the raw data and estimate a model without intercept on these demeaned data.

3.2. Observable versus latent factors

It is usual to consider the idiosyncratic factors, i.e. the components of u in [3.1], as unobservable error terms, but the common factors F are often assumed observable in the literature (see the review of the applied literature in section 3.5). This practice facilitates the estimation, which then disregards the factor dynamics. However, this practice has to be avoided in contagion analysis for the following reasons: (1) the standard literature never confirms the exogeneity of the observable factors and when they are not exogenous, the estimated contagion matrix is biased; (2) it is also necessary to estimate the common factor dynamics in order to predict the future risks; (3) the assumption of latent common factors is mandatory in the financial regulation to capture the uncertainty on risk dependence, to take this dependence into account when computing the required capitals, or when pricing derivatives (see [BCB 10]). Let us discuss the two approaches from both estimation and prediction perspectives, while distinguishing the semi-parametric and parametric approaches of the problem.

3.2.1. *Observable factors*

The inference on the linear dynamic factor model is especially simple, when the systematic factors are observable. Let us denote them $F \equiv X$.

3.2.1.1. *Second-order methods*

Let us assume that the different innovations in u and v are zero mean. In this case, we estimate the parameters of model [3.1] by ordinary least squares (OLS). More precisely:

$$Y_t = BX_t + CY_{t-1} + u_t \qquad [3.6]$$

is a seemingly unrelated regression (SUR) model (see section 3.7.2). Then parameters in B and C matrices can be estimated by OLS performed separately for each row of the system. When the number T of observations is large, the OLS estimators are consistent asymptotically normal. Then consistent approximations of the specific factors are the residuals:

$$\hat{u}_t = Y_t - \hat{B}X_t - \hat{C}Y_{t-1}. \qquad [3.7]$$

In a similar way, the A matrix can be estimated consistently by applying the OLS method to each row of the SUR model:

$$X_t = AX_{t-1} + v_t.$$

Then the innovations v_t are consistently estimated by the OLS residuals:

$$\hat{v}_t = X_t - \hat{A}X_{t-1}. \qquad [3.8]$$

The joint analysis of residuals $\hat{u}_t$ and $\hat{v}_t$ is especially important in our framework. Indeed, to disentangle systematic factor and contagion, we emphasize the role of the exogeneity assumption. Therefore, we have to check if this assumption is satisfied, that is, if the series of error terms u_t and v_t are independent. Information on dependencies can be derived from the cross autocorrelation functions (cross ACF) between the residual series $\hat{u}_{1,t}, ..., \hat{u}_{n,t}, \hat{v}_{1,t}, ..., \hat{v}_{K,t}$ (and also the cross autocorrelation functions of their squares). Indeed we expect zero cross autocorrelations between the $\hat{u}_{i,t}$'s and the $\hat{v}_{k,t}$'s (but also between the different $\hat{u}_{i,t}$'s).

3.2.1.2. *Maximum likelihood method*

When the common distribution of the u_t (respectively, v_t) is specified parametrically, we can improve the efficiency of the estimation method by applying the maximum likelihood (ML) approach. Because of the exogeneity assumption, the optimization of the log-likelihood can be performed separately for the two subsystems corresponding to the factor dynamics and to the dynamics of Y given the factor path, respectively.

Let us denote by $g(u;\alpha)$ and $h(v;\beta)$ the joint probability density functions (p.d.f) of the u_t's and the v_t's, respectively. The objective function to be optimized is:

$$\log L(\theta) = \sum_{t=1}^{T} \log l(y_t, x_t | \underline{y_t}, \underline{x_t}; \theta),$$

where $\underline{y_t}$ (respectively, $\underline{x_t}$) denotes the information on past values of y (respectively, x) and θ the parameters to be estimated: $\theta = (A, B, C, \alpha, \beta)$. By the Bayes formula, we obtain:

$$\begin{aligned}\log L(\theta) &= \sum_{t=1}^{T} \log l(y_t|\underline{y_{t-1}}, \underline{x_t}; \theta) + \sum_{t=1}^{T} \log l(x_t|\underline{y_{t-1}}, \underline{x_{t-1}}; \theta) \\ &= \sum_{t=1}^{T} \log g(y_t - Bx_t - Cy_{t-1}; \alpha) \\ &\quad + \sum_{t=1}^{T} \log h(x_t - Ax_{t-1}; \beta).\end{aligned} \qquad [3.9]$$

Thus, the log-likelihood function can be decomposed as the sum of two partial likelihood functions, which depend on different subsets of parameters. The first component is the log-likelihood of the model for (Y_t) given the systematic factor path, whereas the second component is the log-likelihood associated with the factor dynamics. The ML estimators are obtained by solving the two following optimization problems:

$$\begin{aligned}\left(\hat{B}, \hat{C}, \hat{\alpha}\right) &= \underset{B,C,\alpha}{\arg\max} \textstyle\sum_{t=1}^{T} \log g(y_t - Bx_t - Cy_{t-1}; \alpha), \\ \left(\hat{A}, \hat{\beta}\right) &= \underset{A,\beta}{\arg\max} \textstyle\sum_{t=1}^{T} \log h(x_t - Ax_{t-1}; \beta).\end{aligned} \qquad [3.10]$$

The possibility to separate the two optimization problems implies the asymptotic independence between the ML estimators of matrices B and C and the ML estimator of matrix A.

When the distributions of the error terms are Gaussian, the ML estimators of these matrices are equal to the OLS estimators discussed in the section above, and these OLS estimators are asymptotically efficient.

3.2.2. *Latent common factors*

Let us now consider the case of unobservable factors. As above, we have to distinguish the semi-parametric and parametric approaches.

3.2.2.1. *Second-order methods*

When the common factors are not observable, we have less information and some parameters can not be identified. For instance, we get an observationally equivalent model by changing C into CQ and F_t into $Q^{-1}F_t$, where Q is any invertible (K, K) matrix. Thus, *the latent factors can be identified up to a linear invertible transformation.* To solve this identification problem, we can impose that the variance–covariance matrix of the error term v_t is equal to the identity:

$$V(v_t) = Id_K. \qquad [3.11]$$

Model [3.1] is a vector autoregressive (VAR) model with partial observability. This model admits a state space representation. More precisely, let us introduce the state variable $Z_t = (Y_t', F_t')'$. Under the identification restriction [3.11], we have:

State equation:

$$Z_t = \Psi Z_{t-1} + w_t,$$

where

$$\Psi = \begin{pmatrix} C & BA \\ 0 & A \end{pmatrix}, w_t = \begin{pmatrix} u_t + Bv_t \\ v_t \end{pmatrix}, V(w_t) = \begin{pmatrix} \Omega + BB' & B \\ B' & Id_K \end{pmatrix}.$$

where $\Omega = V(u_t)$.

Measurement equation:

$$Y_t = (Id, 0)Z_t.$$

Consistent estimators of matrices A, B, C and associated variances of the specific factors can be obtained by maximizing a Gaussian pseudo-likelihood[2]. Moreover, the values of this pseudo-likelihood are easily derived numerically by applying the Kalman filter (see section 3.7.3). The linear Kalman filter will also provide linearly filtered values for the factors F_t's.

By considering the VAR representation, we can deduce the first- and second-order moments of the observations (see section 3.7.4).

3.2.2.2. *Maximum likelihood method*

The nonobservability of the underlying factors creates difficulties when applying the ML approach. The identifiability of matrix C and of the factors has to be studied case by case. In general, the log-likelihood function cannot be decomposed as a sum of partial log-likelihoods as in the case of observable factors. It admits a complicated expression, since the unobservable factor path has to be integrated out. This log-likelihood becomes:

$$\log L = \log \left[\int .... \int \prod_{t=1}^{T} g(y_t - Bf_t - Cy_{t-1}; \alpha) \times \prod_{t=1}^{T} h(f_t - Af_{t-1}; \beta) \prod_{t=1}^{T} df_t \right], \qquad [3.12]$$

where $df_t = df_{1t} \dots df_{Kt}$.

It involves an integral of a large dimension KT, which increases with the number of observation dates. Even if the theoretical asymptotic properties of the ML estimators are known, their estimates can be difficult to derive. Indeed the

2 That is a likelihood function computed under the assumption of Gaussian errors. The associated maximum likelihood estimators are consistent even if the true distribution of the errors is not Gaussian. This explain the term “pseudo”.

optimization of log-likelihood [3.12] can be numerically too demanding, and it can be preferable to consider alternative estimation approaches numerically simpler and implying a rather small loss of efficiency. Simulation-based approaches or method of moments based on Laplace transforms are of this type and are used, for instance, in the application of Chapter 6.

3.3. Shocks, impulse response functions and stress

Let us consider dynamic model [3.1]. This specification involves two types of innovations, that are the innovations v_t for the common factor dynamics and the innovations u_t for the dynamic of the observed variables given the common factor path. The model allows for two types of shocks either on the components of the specific factors, or on the innovations of the systematic factor. Let us discuss the consequences of such shocks at different maturities.

3.3.1. *Shock on specific factors*

Let us shock the specific factor u at a given date, say, $t = 0$, and denote δu as the magnitude of the shock. Since the common factor is exogenous, there is no impact on the dynamics of F. However, there is an impact on the whole future trajectory of Y. This impact is such that:

$$\delta Y_t = C\delta Y_{t-1}, t \geq 1, \qquad [3.13]$$

with the initial condition: $\delta Y_0 = \delta u$.

We deduce the impulse response function, that is the evolution of this effect in terms of the time-to-maturity and shock magnitude:

$$\delta Y_t = C^t \delta u. \qquad [3.14]$$

The impulse response function depends on the initial shock in a multiplicative way, which is a standard feature of such linear systems (see e.g. [KOO 96, GOU 05]). The pattern of the impulse response function depends on the eigenvalues of matrix C only. Under the stability condition, the impulse response function tends to zero in the long run. Its pattern depends on the fact that C is diagonalizable, or simply triangularizable, and on the type of eigenvalues of matrix C (see the discussion in section 2.1.2). We obtain a superposition of different patterns: exponentially decreasing components corresponding to the positive real eigenvalues, sinusoidal amortizing components of various periodicities for the negative real or complex eigenvalues, and possibly such patterns multiplied by polynomials when C cannot be diagonalized. This is the well-known resonance effect (see Figure 3.1).

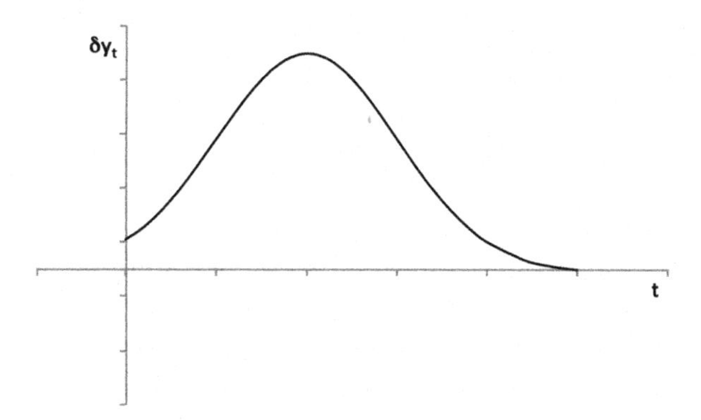

Figure 3.1. *IRF with resonance*

The pattern associated with resonance is especially important for economic or financial policies. It means that the maximal effect of the shock is not immediate, but occurs with a lag depending on the location of the bump.

3.3.2. *Shock on the systematic factor*

When the shock is introduced at date, say, $t = 0$, by means of the innovation v_0 with magnitude δv, both factor and endogenous variable dynamics will change. The equations defining the impact of the shocks are:

$$\delta Y_t = B\delta F_t + C\delta Y_{t-1} \text{ and } \delta F_t = A\delta F_{t-1}, t \geq 1, \qquad [3.15]$$

with initial conditions : $\delta Y_0 = 0$, $\delta F_0 = \delta v$.

We deduce:

$$\delta F_t = A^t \delta v,$$

and by substitution:

$$\delta Y_t = BA^t \delta v + C\delta Y_{t-1}.$$

Finally, we get the impulse response function:

$$\delta Y_t = [BA^t + CBA^{t-1} + ... + C^{t-1}BA]\delta v. \qquad [3.16]$$

As for the shocks on specific factors, the impact on the response function is proportional to the magnitude of the initial shock. But now the pattern of the impulse response function depends on both the eigenvalues of matrices A and C, and also on the possibility to diagonalize, or not, matrix A. Finally, the expression in equation [3.16] shows the direct effect of the shock equal to $BA^t\delta v$, and the additional effect due to contagion equal to $[CBA^{t-1} + ... + C^{t-1}BA]\delta v$.

3.3.3. *Stress tests*

The recent regulations for Finance and Insurance, such as Basel 3, Solvency 2 (see (BSB 10]), and the supervision for Financial Stability demand to perform stress tests to compute the required capital for systemic risk, or to be used to rank the financial institutions and detect the Systemically Important Financial Institutions (SIFI) (see [FIN 13]).

Such stress tests compare the "standard situation" or the current situation to "stressed situations". Typically, a standard situation corresponds to the dynamic model [3.1] estimated on a calm, quiet period. Then the additional risks, due to stress on systemic factors, are analyzed through the patterns and magnitudes of the impulse response functions following shocks on the systematic factors.

How do we choose these shocks? Let us assume a single factor, $K = 1$, and an analysis performed at date, say T. Different factor values are relevant, that are the mean value, equal to zero in our case, which corresponds to the average (standard) situation, the current factor value F_T, which can be proxied by its filtered value $\hat{F}_T$, and some extreme values of the factor corresponding to 5% and 95% percentiles of the conditional distribution of F_T given $\hat{F}_{T-1}$, for instance. Both extreme percentiles have to be considered since some assets might depend negatively on the factor value, when they partly ensure against systematic risk.

Let us consider the stock returns of a set of banks. The associated impulse response functions can be used to fix the required capital for systemic risk or to define ratings for systemic risk from stock market data. At date T, the additional reserve for systemic risk can be deduced from the difference between the impulses in one extreme stress and in the current situation.

This can be done for the different stocks, i.e. banks. Let us consider Figure 3.2 with the difference displayed for two banks. We immediately note that these differences depend on the short-, medium- and long-term horizons, and can cross together. Thus, the level of additional required capital and the respective ranking of these two banks for systemic risk will also depend on the selected horizon. We observe that bank 2 is more sensitive to systemic risk than bank 1 in the short run, and the reverse in the long run.

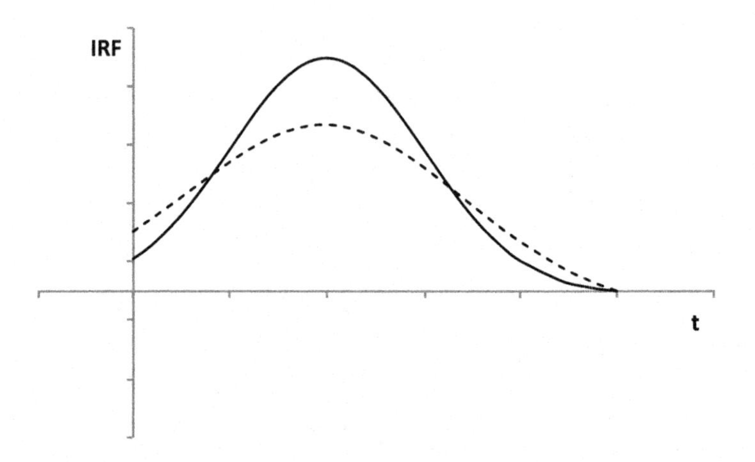

Figure 3.2. *The difference between impulse responses. This figure displays the IRF for bank 1 (continuous line) and the IRF for bank 2 (dotted line)*

3.4. Constrained models and misspecification

Usually, the literature considers constrained versions of the dynamic linear model [3.1] by either introducing systematic factor only, or contagion only. If model [3.1] is the right one, this will induce misspecifications, with severe consequences on the estimated parameters and on the decisions based on these parameters, concerning the required capital, the portfolio management, or the financial stability policy. We discuss below the existence and magnitude of the biases.

3.4.1. *Pure contagion pseudo-model*

This corresponds[3] to the case of no systematic factor ($B = 0$).

– Let us first assume that the model is estimated by OLS under its pure contagion version as:

$$Y_t = C^* Y_{t-1} + u_t^*. \quad [3.17]$$

The estimated contagion matrix will converge to the theoretical autoregressive matrix coefficient when regressing Y_t on Y_{t-1}. Thus, we obtain:

$$C^* = Cov(Y_t, Y_{t-1}) V(Y_{t-1})^{-1}, \quad [3.18]$$

where the expressions of the variance and of the autocovariance at order 1 are derived from the Yule–Walker equations written on the vector form [3.3] (see section 3.7.5). We obtain:

$$C^* = C + BA\ Cov(F_{t-1}, Y_{t-1}) V(Y_{t-1})^{-1}. \quad [3.19]$$

Matrix C^* is different from the true contagion matrix C, whenever $BA\ Cov(F_{t-1}, Y_{t-1})\ V(Y_{t-1})^{-1}$ is different from 0. In fact, C^* indirectly captures a part of the omitted systematic factor effects. This will change the form of the network associated with the contagion matrix and generally increase the ideas about the magnitudes of the contagion effects.

To illustrate these effects, let us consider the following bidimensional model with a single factor:

$$\begin{pmatrix} Y_{1,t} \\ Y_{2,t} \end{pmatrix} = \begin{pmatrix} b_1 \\ b_2 \end{pmatrix} f_t + \begin{pmatrix} 0 & c_{12} \\ 0 & 0 \end{pmatrix} \begin{pmatrix} Y_{1,t-1} \\ Y_{2,t-1} \end{pmatrix} + \begin{pmatrix} u_{1,t} \\ u_{2,t} \end{pmatrix}, \quad [3.20]$$

3 A pseudo-model is a misspecified model.

with $f_t = af_{t-1} + v_t$, $Eu_t = 0, Ev_t = 0, Vu_t = \sigma^2 Id, Vv_t = Id$.

Matrix C shows a unidirectional contagion from 2 to 1. The expression of the matrix C^* is derived in section 3.7.6. We obtain:

$$C^* = \begin{pmatrix} 0 & c_{1,2} \\ 0 & 0 \end{pmatrix} + a\left(\frac{a^2}{1-a^2} + \sigma^2\right)\begin{pmatrix} b_1 \\ b_2 \end{pmatrix}$$

$$\times (b_1 + c_{12}b_2, b_2)\, V \begin{pmatrix} Y_{1,t} \\ Y_{2,t} \end{pmatrix}^{-1}. \qquad [3.21]$$

Thus, in general, the misleading contagion matrix C^* has non-zero elements everywhere.

– Let us now assume that the pure contagion model is estimated by OLS under its extended version including more lags:

$$Y_t = C^*(L)Y_{t-1} + u_t^*. \qquad [3.22]$$

It is known that the process Y_t following dynamics [3.1] admits a weak autoregressive representation as [3.22] with a lag polynomial of degree $p_C = \infty$, and error terms which are uncorrelated, but may be dependent. Thus, the omission of the systematic factor in the analysis will give the misleading impression of a longer memory.

3.4.2. *Pure factor pseudo-model*

When no contagion is introduced ($C = 0$), the model becomes:

$$Y_t = B^* F_t^* + u_t^*, \text{ with } F_t^* = A^* F_{t-1}^* + v_t^*. \qquad [3.23]$$

The first subsystem is the specification of exploratory factor analysis, but an exogenous autoregressive dynamics is now assumed for the factor. The asymptotic biases, when estimating matrices B^* and A^*, depend on the estimation method used and are difficult to derive. But intuitively, model [3.1] involves a large number $n + K$ of regressors, that are the components of F_t and the components of Y_{t-1}. In practice, we expect the estimated number K^* of regressors in pseudo-model [3.16] to be strictly larger than K. Thus, we will have an overestimation of the number of systematic factors, including, most likely, the main factors F.

This remark is often used in practice to get an idea of the number K of factors in model [3.1]. The econometrician applies a standard (static) exploratory factor analysis on the data Y_t, which will provide an upper bound K^* for the true number of factors.

3.5. The literature

The dynamic linear model [3.1], or its extensions to include more lags (see section 3.1.2) are used in finance to analyze the joint dynamics of either asset returns in a same market, index returns for different markets or interest rates of various time-to-maturities.

3.5.1. *Joint analysis of market returns*

From the 1980s until now, the international stock markets have been increasingly influenced by globalization. Even if most studies argue that the United States market has a leading character, these analysis strongly depend on the period of observation, on the selected market indices and on the type of linear dynamic model used for the analysis. To analyze the transmission between markets, the studies generally consider a model with contagion, but without common factors, and apply Granger causality measures to determine the type of contagion between markets (see section 3.7.1). For instance, Malliaris and Urriata [MAL 92] applied this methodology to six market indices. They found no lead-lag relationship before and after the 1987 crisis, but feedback relationships during the crisis, whereas Arshanapalli and Doukas [ARS 93] claim that, since this crash, a lot of market indices are driven by the United States. Using a same type of methodology on a more recent period, Balios and Xanthakis [BAL 03] also observed that the Dow Jones Industrial Average (DGIA) index is the leading index for the main European (and Japanese) indices, but, if the contagion is direct with the British index, the Footsie, it can pass through the Footsie for other European indices as the French CAC 40.

However, such analyses can be misleading since they omit the possible common factors. They also disregard the stochastic volatility and covolatility effects, as well as the transmission of risks by means of risk premia.

3.5.2. *Term structure of interest rates*

Let us focus on the risk-free interest rates. An interest rate $r(t,h)$ is indexed by time t and term (or time-to-maturity) h. The short-term interest rate corresponds to $h = 1$, whereas long-term interest rates correspond to large values of h. More generally the observations available at time t: $Y_t = [r(t,1),\ldots,r(t,H)]'$ concern the curve $h \to r(t,h)$, $h = 1,\ldots,H$. This curve is called the term structure of interest rates at date t. Thus, the dynamic models are introduced to understand the evolution of such curves. The basic models are pure factor models, called affine term structure models (ATSM) [DUF 96, GOU 06]:

$$r(t,h) = \alpha(h) + b(h)F_t + \epsilon(t,h), h = 1,\ldots,H, t = 1,\ldots,T. \quad [3.24]$$

Thus, up to the errors ϵ, the term structure is written as a combination of baseline curves, that are $\alpha(.)$ and the components of $b(.)$, with stochastic weights, that are the components of F_t.

The standard specifications introduce three unobserved factors. The idea is to get enough flexibility to allow for different consequences of shocks on F on the pattern of the term structure. They are known as a change of level, a change of slope and a change of curvature.

These ideas are illustrated in Figure 3.3, with a flat term structure $r(t, h) = r_{0,t}$ independent of h, as the benchmark, and the various shocked term structures.

Once the parameters, that are the sensitivities α, β, and the factor dynamics, have been estimated, we can approximate the factors by minimizing:

$$\hat{F}_t = \arg\min_{F_t} \sum_{h=1}^{H} \left[r(t, h) - \hat{\alpha}(h) - \hat{\beta}(h) F_t \right]^2 . \qquad [3.25]$$

The solutions $\hat{F}_{k,t}$, $k = 1, 2, 3$, are linear combinations of the interest rates:

$$\hat{F}_{k,t} = \sum_{h=1}^{H} c_k(h) r(t, h), \text{ say.} \qquad [3.26]$$

They can be interpreted as the rate of a coupon bond with coupon $c_k(h)$ for term h, $h = 1, \ldots, H$. Up to some linear transformations of the factors (remind that they are defined up to an invertible linear transformation), it is expected to find:

– a first factor $\hat{F}_{1,t} \approx \sum_{h=1}^{H} r(t, h)$, with almost equal weights. It is interpretable as a bond index and is very sensitive to the level effect.

– a second factor $\hat{F}_{2,t} \approx -r(t, 1) + r(t, H)$, measuring a spread between long-term and short-term interest rates. The associated coupon bond is weakly sensitive to the level effect, but very sensitive to the slope effect.

– a third factor $\hat{F}_{3,t} \approx r(t,1) - 2r(t,H^*) + r(t,H)$, where $H^* \approx H/2$. This "butterfly" bond will be very sensitive to the curvature effect and weakly sensitive to both the level and slope shocks.

Lagged values have been introduced rather recently in affine term structure models:

$$Y_t = \alpha + BF_t + CY_{t-1} + \epsilon_t, \text{ say,} \qquad [3.27]$$

with, clearly, the curse of dimensionality if the number H of terms is large. In such extended models, the matrix C has to be constrained. Due to the standard interpretation in terms of level, slope and curvature, it seems interesting to look for a matrix C with rank three, related with the structure of coupons of the mimicking bonds. In other words, the constrained model becomes:

$$Y_t = \alpha + BF_t + B^*\tilde{F}_t + \epsilon_t, \text{ say,} \qquad [3.28]$$

with both the true latent factor F_t and their market proxies $\tilde{F}_t$.

3.6. Chapter 3 highlights

The effect of common factors and the contagion effects can be identified in linear dynamic models. We explain how to estimate these effects by either parametric or semi-parametric approaches, distinguishing the case of observable factors and of latent factors. We also discussed the misleading interpretation of the result when either the frailty, or the contagion is omitted.

Finally, we illustrate the use of such models in the literature for the joint analysis of market indices, and for understanding the term structure of interest rates and its evolution.

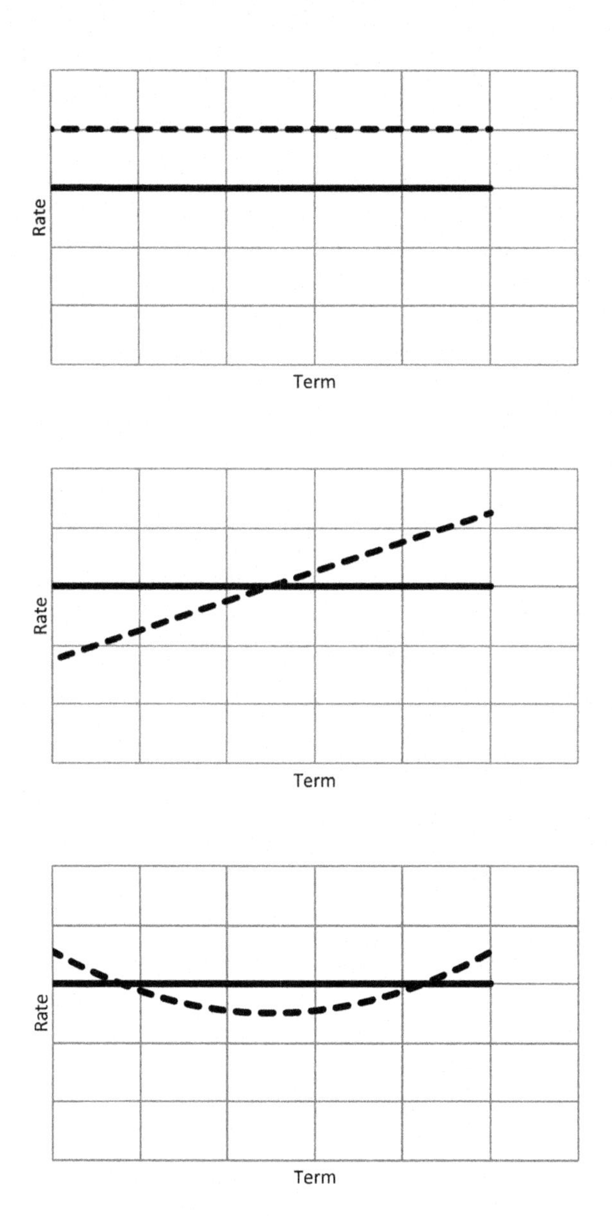

Figure 3.3. *Level, slope and curvature. This figure displays a flat term structure as the benchmark (continuous line), and various shocked term structures (dotted line), when the shock concerns the level (top), the slope (middle), and the curvature (bottom)*

3.7. Appendices

3.7.1. *Granger causality*

The notion of causality can be defined in a linear or in a nonlinear framework. We introduce below the definition in a nonlinear framework, the linear definitions corresponding to the special case of Gaussian distributions. Let us consider a joint Markov process $(X_t, Y_t)'$ with transition p.d.f. denoted by: $l(y_t, x_t|y_{t-1}, x_{t-1})$.

By Bayes formula, this joint conditional p.d.f. can be decomposed as follows:

$$l(y_t, x_t|y_{t-1}, x_{t-1}) = l(y_t|y_{t-1}, x_t, x_{t-1})l(x_t|y_{t-1}, x_{t-1}) \qquad [3.29]$$

$$= l(x_t|y_t, y_{t-1}, x_{t-1})l(y_t|y_{t-1}, x_{t-1}). \qquad [3.30]$$

DEFINITION 3.1.– *Under the joint Markov assumption,*

1) X_t *does not Granger cause* Y_t *if*

$$l(y_t|y_{t-1}, x_{t-1}) = l(y_t|y_{t-1}).$$

2) Y_t *does not Granger cause* X_t *if*

$$l(x_t|y_{t-1}, x_{t-1}) = l(x_t|x_{t-1}).$$

3) Y_t *and* X_t *are conditionally independent if*

$$l(y_t, x_t|y_{t-1}, x_{t-1}) = l(y_t|y_{t-1}, x_{t-1})l(x_t|y_{t-1}, x_{t-1})$$

$$\Leftrightarrow l(y_t|y_{t-1}, x_t, x_{t-1}) = l(y_t|y_{t-1}, x_{t-1})$$

$$\Leftrightarrow l(x_t|y_t, y_{t-1}, x_{t-1}) = l(x_t|y_{t-1}, x_{t-1}).$$

When Y_t does not cause X_t, the process (X_t) has an autonomous dynamic. It is often said that this process (X_t) is exogenous (with respect to (Y_t)).

By looking at the different non causalities, we try to structure and simplify the analysis of serial dependencies between processes (X_t) and (Y_t). In particular, we get the following characterization of the independence between both processes.

PROPOSITION 3.1.– The processes (X_t) and (Y_t) are independent if:

1) (Y_t) *and* (X_t) *are conditionally independent;*

2) (Y_t) *does not Granger cause* (X_t)*;*

3) (X_t) *does not Granger cause* (Y_t).

3.7.2. *Seemingly unrelated regression (SUR) model*

Let us consider a multivariate regression model with the same explanatory variables in all equations and possibly correlated errors:

$$y_{i,t} = x_t b_i + \epsilon_{i,t}, \quad i = 1, \ldots, n, \quad t = 1, \ldots, T, \qquad [3.31]$$

where the errors are i.i.d. Gaussian:

$$\epsilon_t \sim N(0, \Omega). \qquad [3.32]$$

For a given variance–covariance matrix Ω, the Gaussian maximum likelihood estimator of the regression coefficients b_i, $i = 1, ..., n$, coincide with their generalized least squares (GLS) estimators. They are solutions of the minimization problem:

$$\hat{B} \equiv (\hat{b}_1, ..., \hat{b}_n) = \arg \min_{b_1,\ldots,b_n} \sum_{t=1}^{T} (y_{1,t} - x_t b_1, ..., y_{n,t} - x_t b_n) \Omega^{-1}$$
$$\times (y_{1,t} - x_t b_1, ..., y_{n,t} - x_t b_n)'.$$

Let us denote $w^{i,j}$ as the generic element of the matrix Ω^{-1}. we have:

$$\hat{B} = \arg \min_{b_1,\dots,b_n} \sum_{i=1}^{n} \sum_{j=1}^{n} \left\{ w^{i,j} \sum_{t=1}^{T} (y_{i,t} - x_t b_i)(y_{j,t} - x_t b_j) \right\}.$$

The first-order conditions are:

$$\sum_{j=1}^{n} w^{i,j} \sum_{t=1}^{T} x_t'(y_{j,t} - x_t b_j) + w^{i,j} \sum_{t=1}^{T} x_t'(y_{i,t} - x_t b_i) = 0,$$

$$i = 1, \dots, n. \qquad [3.33]$$

We see that the OLS estimators, which satisfy the equations:

$$\sum_{t=1}^{T} x_t'(y_{i,t} - x_t \hat{b}_i) = 0, \quad i = 1, \dots, n, \qquad [3.34]$$

satisfy the first-order conditions above; we deduce the following as a result:

PROPOSITION 3.2.– In the multivariate model [3.31] with the same explanatory variables in all equations, the GLS estimators of regression coefficients are equal to the OLS estimators computed equation by equation.

This justifies the terminology *seemingly unrelated regression* (SUR) of model [3.31]. Indeed, we can estimate the b_j's as if the errors were unrelated: $\Omega = Id$. The result in proposition 3.2 can in particular be used for a VAR(1) model where $X_t = Y_{t-1}$.

3.7.3. *Kalman filter*

The Kalman filter is an algorithm to update the predictions, and their accuracies, in linear dynamic models with an unobserved factor (see e.g. [GOU 97]).

i) *State space model*

In their simplest version, these dynamic models have the following structure:

$$F_t = AF_{t-1} + \epsilon_t, \qquad [3.35]$$

$$Y_t = CF_t + u_t, \qquad [3.36]$$

where u_t nd ϵ_t are two independent Gaussian noises with zero mean and variance–covariance matrices:

$$V(u_t) = Q, V(\epsilon_t) = R.$$

The factor F_t is often called the state variable and the equation defining its dynamics [3.36] called the state equation. The observed variable Y_t is called the measurement and equation [3.35] is the measurement equation.

ii) *The filter*

The filter provides recursive equations to compute:

– the prediction ${}_t\hat{F}_t$ of factor F_t given the observations $\underline{Y_t} = (Y_t, Y_{t-1}, ...)$;

– the accuracy of these predictions measured by ${}_t\Sigma_t = V(F_t - {}_t\hat{F}_t)$.

These equations have to be used jointly. We have:

$${}_t\hat{F}_t = A\,{}_{t-1}\hat{F}_{t-1} + K_t\left(Y_t - CA\,{}_{t-1}\hat{F}_{t-1}\right), \qquad [3.37]$$

$${}_t\Sigma_t = (Id - K_tC)\left(A\,{}_{t-1}\Sigma_{t-1}A' + R\right), \qquad [3.38]$$

where

$$K_t = \left[A\,{}_{t-1}\Sigma_{t-1}A' + R\right]C' \times \left[C(A\,{}_{t-1}\Sigma_{t-1}A' + R)C' + Q\right]^{-1}. \qquad [3.39]$$

Thus, at date t, we first compute the matrix K_t from ${}_{t-1}\Sigma_{t-1}$. K_t is called the gain of the filter. Then, we deduce ${}_t\Sigma_t$ from ${}_{t-1}\Sigma_{t-1}$ by equation [3.38], and ${}_t\hat{F}_t$ from ${}_{t-1}\hat{F}_{t-1}$, by equation [3.37]. The filter can be completed to get the predictions of the observable variables: ${}_{t-1}\hat{Y}_t = E(Y_t|\underline{Y_{t-1}})$ (and their accuracy). For instance, we have:

$$ {}_{t-1}\hat{Y}_t = CA\, {}_{t-1}\hat{F}_{t-1}. \qquad [3.40] $$

iii) *The Gaussian likelihood function*

Due to the nonobservability of the factor path, the Gaussian likelihood corresponding to the state space model [3.36]-[3.35] has a complicated expression. Indeed the unobservable factor path has to be integrated out; therefore, the expression of the likelihood involves multidimensional integrals of dimension TK, where $K = \dim F_t$. However, it is known that the Gaussian likelihood is also equal to:

$$ \begin{aligned} l(y_1, ..., y_T) &\simeq \prod_{t=1}^{T} l(y_t|\underline{y_{t-1}}) \\ &= \prod_{t=1}^{T} \left[\frac{1}{(2\pi)^{n/2}} \frac{1}{\sqrt{\det {}_{t-1}\Omega_t}} \right. \\ &\quad \left. \times \exp\left[-\frac{1}{2}(Y_t - {}_{t-1}\hat{Y}_t)\, {}_{t-1}\Omega_t^{-1}(Y_t - {}_{t-1}\hat{Y}_t)' \right] \right], \end{aligned} $$

where: ${}_{t-1}\hat{Y}_t = E(Y_t|\underline{Y_{t-1}})$, ${}_{t-1}\Omega_t^{-1} = V(Y_t - {}_{t-1}\hat{Y}_t)$.

Since the Kalman filter is an algorithm to recursively compute the sequences $t \rightarrow {}_{t-1}\hat{Y}_t$, $t \rightarrow {}_{t-1}\Omega_t$ for any given set of parameters A, C, Q, R, this filter also provides the numerical computation of the likelihood function. Thus, the ML estimates can be derived.

3.7.4. *Second-order properties*

By considering the VAR representation, we deduce its first- and second-order moments. We obtain:

$$\begin{cases} E(Z_t) & = 0, \\ V(Z_t) & = \Gamma_Z(0) = V(w) + \Psi V(w)\Psi' + \\ & \qquad \dots + \Psi^h V(w)\Psi^{h'} + \dots, \\ Cov(Z_t, Z_{t-h}) & = \Gamma_Z(h) = A^h \Gamma_Z(0), h \geq 0. \end{cases}$$

These expressions can be used to derive the second-order properties of the process (Y_t).

LEMMA 3.1.– We have:

$$\Psi^h = \begin{pmatrix} C^h & D_h \\ 0 & A^h \end{pmatrix},$$

where $D_h = CD_{h-1} + BA^h, h \geq 1, D_0 = 0$, or equivalently $D_h = \sum_{k=0}^{h-1}(C^k BA^{h-k})$.

PROOF 3.1.– Since:

$$\Psi^h = \Psi\Psi^{h-1} = \begin{pmatrix} C & BA \\ 0 & A \end{pmatrix} \begin{pmatrix} C^{h-1} & D_{h-1} \\ 0 & A^{h-1} \end{pmatrix},$$

we deduce $D_h = CD_{h-1} + BA^h$. □

LEMMA 3.2.– We have:

$$V(Z) = \begin{bmatrix} V(Y) & Cov(Y,F) \\ Cov(F,Y) & V(F) \end{bmatrix},$$

where:

$$V(F) = AV(F)A' + \Omega = \sum_{h=0}^{\infty} A^h \Omega A^{h'},$$

$$Cov(Y,F) = CCov(Y,F)A' + BAV(F)A' + B\Omega,$$

$$V(Y) = \sum_{h=0}^{\infty} C^h \Delta C^{h'},$$

with

$$\Delta = BACov(F,Y)C' + CCov(Y,F)A'B' \\ +BAV(F)A'B' + \Sigma + B\Omega B'.$$

PROOF 3.2.– We have:

$$\Gamma_Z(0) = \Psi\Gamma_Z(0)\Psi' + V(w)$$

$$= \begin{pmatrix} C & BA \\ 0 & A \end{pmatrix} \begin{pmatrix} V(Y) & Cov(Y,F) \\ Cov(F,Y) & V(F) \end{pmatrix} \begin{pmatrix} C' & 0 \\ A'B' & A' \end{pmatrix}$$

$$+ \begin{pmatrix} \Sigma + B\Omega B' & B\Omega \\ \Omega B' & \Omega \end{pmatrix}.$$

By identification, we deduce the recursive system:

$$\begin{aligned} V(F) &= AV(F)A' + \Omega, \\ Cov(Y,F) &= CCov(Y,F)A' + BAV(F)A' + B\Omega, \\ V(Y) &= CV(Y)C' + BACov(F,Y)C' + CCov(Y,F)A'B' \\ &\quad + BAV(F)A'B' + \Sigma + B\Omega B'. \end{aligned}$$

□

These lemmas are used to deduce equations satisfied by the autocovariance function of the process (Y_t). Indeed, we have:

$$Cov(Z_t, Z_{t-h}) = \Psi^h \Gamma_Z(0) = \begin{pmatrix} C^h & D_h \\ 0 & A^h \end{pmatrix} \begin{pmatrix} V(Y) & Cov(Y,F) \\ Cov(F,Y) & V(F) \end{pmatrix}.$$

By considering the first block-diagonal element of this product, we get the following result:

PROPOSITION 3.3.– The second-order properties of (Y_t) are the following:

i) the autocovariance function $\Gamma_Y(h)$ of the observable process Y is equal to:

$$\Gamma_Y(h) = Cov(Y_t, Y_{t-h}) = C^h \Gamma_Y(0) + D_h Cov(F, Y), h \geq 0,$$

where $D_h = \sum_{k=0}^{h-1}(C^k B A^{h-k}) = C D_{h-1} + B A^h$;

ii) the unconditional covariance between the observable process Y and the frailty $Cov(Y, F)$ is the solution of:

$$Cov(Y, F) = C Cov(Y, F) A' + BAV(F)A' + B\Omega.$$

We get the standard component $C^h \Gamma_Y(0)$ for a VAR dynamic of the observable process (Y_t) with autoregressive matrix C plus the term $D_h Cov(F, Y)$ due to the unobservable factor. D_h is a rather complicated function of h. Indeed, the unobservability of F implies a $VAR(\infty)$ dynamics with an infinite autoregressive lag, when process (Y_t) is considered alone.

The unconditional covariance between Y and F is the solution of a system of Riccati equations, which in general has to be solved numerically. The different formulas are greatly simplified for a single factor model.

3.7.5. *The Yule–Walker equation of order 1*

Let us consider the autocovariance at order 1. We obtain:

$$Cov\left(\begin{pmatrix} Y_t \\ F_t \end{pmatrix}, \begin{pmatrix} Y_{t-1} \\ F_{t-1} \end{pmatrix}\right) = \begin{pmatrix} C & BA \\ 0 & A \end{pmatrix} V \begin{pmatrix} Y_{t-1} \\ F_{t-1} \end{pmatrix},$$

since the joint process $(Y_t, F_t)'$ is stationary. In particular, by considering the North-West block we obtain:

$$Cov(Y_t, Y_{t-1}) = C\ V(Y_{t-1}) + BA\ Cov(F_{t-1}, Y_{t-1}),$$

or equivalently,

$$\begin{aligned} C^* &= Cov(Y_t, Y_{t-1})V(Y_{t-1})^{-1} \\ &= C + BA\ Cov(F_{t-1}, Y_{t-1})V(Y_{t-1})^{-1}. \end{aligned}$$

3.7.6. *Misspecified pure contagion model*

Let us apply lemma 3.2 in section 3.7.4. We obtain,

$$Vf = \frac{1}{1-a^2}, \qquad [3.41]$$

and

$$\begin{pmatrix} Cov(y_1, f) \\ Cov(y_2, f) \end{pmatrix} = a \begin{pmatrix} 0 & c_{12} \\ 0 & 0 \end{pmatrix} \begin{pmatrix} Cov(y_1, f) \\ Cov(y_2, f) \end{pmatrix} + \begin{pmatrix} b_1 \\ b_2 \end{pmatrix} \left(\frac{a^2}{1-a^2} + \sigma^2 \right). \qquad [3.42]$$

We deduce:

$$\begin{aligned} Cov(y_2, f) &= b_2 \left(\frac{a^2}{1-a^2} + \sigma^2 \right), \\ Cov(y_1, f) &= (b_1 + c_{12} b_2) \left(\frac{a^2}{1-a^2} + \sigma^2 \right). \end{aligned}$$

4

Applications of Linear Dynamic Models

In this chapter, we explain how the static and dynamic models of Chapter 3 can be used for portfolio management, risk monitoring and the analysis of financial stability.

4.1. Portfolio management

Let us first consider both static and dynamic linear factor models for excess asset returns and analyze the standard mean-variance portfolio management in these two cases. We distinguish the effects of the common factor, contagion and idiosyncratic risks on the efficient portfolio allocations and their Sharpe performances.

4.1.1. *Asset return models*

Let us consider n risky assets and one risk-free asset. We denote $y_{i,t+1}$, $i = 1, ..., n$, the returns on risky assets on period $(t, t+1)$, and $y_{0,t}$ the risk-free return on the same period. The risk-free return for this period is indexed by t, since it is known at the beginning of the period.

4.1.1.1. *Static model*

We assume that the excess asset returns, which are the differences between the risky and risk-free returns $\tilde{y}_{i,t+1} = y_{i,t+1} - y_{0,t}$, satisfy the static one factor model:

$$\tilde{y}_{i,t+1} = \beta_i F_{t+1} + u_{i,t+1}, i = 1, ..., n,$$

or equivalently:

$$\tilde{Y}_{t+1} = \beta F_{t+1} + u_{t+1}, \quad [4.1]$$

where $\beta = (\beta_1, ..., \beta_n)'$ and $u_{t+1} = (u_{1,t+1}, ..., u_{n,t+1})' \sim IIN(0, \sigma^2 Id_n)$. The unobserved latent factor satisfies:

$$F_{t+1} = \mu + v_{t+1}, \; v_{t+1} \sim IIN(0, \eta^2), \quad [4.2]$$

where the errors v_{t+1} and u_{t+1} are independent. The parameter β_i is the sensitivity or beta of excess return $\tilde{y}_{i,t+1}$ to factor F_t.

In Chapter 3, we assumed a zero intercept $\mu = 0$. In this section, we introduce a non-zero factor intercept to remunerate the additional risk exposure. Let us now compute the first-order and second-order moments of the excess returns. Let us assume that the underlying factor is not observed. Thus, we focus on the marginal (i.e. unconditional) moments. The expected excess returns are $E(\tilde{y}_{i,t+1}) = \mu\beta_i, \; i = 1, ..., n$, and the idiosyncratic risks are measured by $V(u_{i,t+1}) = \sigma^2$. They do not depend on the asset.

The systematic source of risk through the common unobserved factor creates both an additional individual risk equal to $\eta^2\beta_i^2$:

$$V(\tilde{y}_{i,t+1}) = \sigma^2 + \eta^2\beta_i^2, i = 1, \ldots, n,$$

and a dependence between the excess returns of two different risky assets, since:

$$Cov(\tilde{y}_{i,t+1}, \tilde{y}_{j,t+1}) = \eta^2 \beta_i \beta_j, \text{ for } i \neq j.$$

These moments can be written under a matrix form, as:

$$E(\tilde{Y}_{t+1}) = \mu\beta, \quad [4.3]$$

$$V(\tilde{Y}_{t+1}) \equiv \Omega = \sigma^2 Id + \eta^2 \beta\beta'. \quad [4.4]$$

This variance-covariance matrix can be decomposed in terms of orthogonal projectors as follows (see section 4.4.1):

$$\Omega = \sigma^2 \left(Id - \frac{\beta\beta'}{\beta'\beta} \right) + \lambda^2 \frac{\beta\beta'}{\beta'\beta}, \quad [4.5]$$

where $\lambda^2 = \sigma^2 + \eta^2 \beta'\beta$.

This decomposition can be used to derive the spectral decomposition (that is the set of eigenvalues and eigenvectors) and the inverse of matrix Ω. Indeed, the variance-covariance matrix Ω admits as eigenvalues σ^2, with multiplicity order $n-1$, and λ^2 with multiplicity order 1. We deduce that (see section 4.4.1):

$$\Omega^{-1} = \frac{1}{\sigma^2} \left(Id - \frac{\beta\beta'}{\beta'\beta} \right) + \frac{1}{\lambda^2} \frac{\beta\beta'}{\beta'\beta}. \quad [4.6]$$

REMARK 4.1.– This static model is easily extended to asset-dependent idiosyncratic risks: the specific errors $u_{i,t}$ are still independent, Gaussian, with zero-mean, but different variances $V(u_{i,t}) = \sigma_i^2$, $i = 1, ..., n$. In this extension, the asset i is characterized by the beta β_i and the specific variance σ_i^2. Similarly, we could also introduce an asset-dependent intercept μ_i in the equation:

$$\tilde{y}_{i,t+1} = \mu_i + \beta_i F_{t+1} + u_{i,t+1}, i = 1, ..., n.$$

Therefore, the asset could be characterized by $(\mu_i, \beta_i, \sigma_i^2)$.

We consider in this section the simplified model [4.1]–[4.2], since some computations are much easier, such as the computation of the inverse matrix Ω^{-1} in [4.6], and also, since the additional intercept $\mu_i = 0$ in standard equilibrium models, such as the capital asset pricing model (CAPM) (see [SHA 64] and [LIN 65]).

4.1.1.2. *Dynamic model*

Thus, dynamic features can be introduced in both the transition equation defining the factor F_t and in the measurement equation. We can now assume:

$$\tilde{y}_{i,t+1} = \beta_i F_{t+1} + \sum_{j=1}^{n} c_{i,j} \tilde{y}_{j,t} + u_{i,t+1}, i = 1, ..., n,$$

where the errors $u_{i,t}$, $i = 1, \ldots, n$ are mutually independent such that: $u_{i,t} \sim N(0, \sigma^2)$, whereas the unobserved latent factor satisfies the autoregressive model:

$$F_{t+1} = \mu + \varphi F_t + v_{t+1}, v_{t+1} \sim N(0, \eta^2).$$

The two sequences of error terms (u_t) and (v_t) are independent. The parameter β_i is the sensitivity, or beta, of return $\tilde{y}_{i,t+1}$ with respect to the unobserved factor, while parameters $c_{i,j}, j = 1, ..., n$ represent the effects of lagged asset returns. The associated matrix representation is:

$$\begin{cases} \tilde{Y}_{t+1} = \beta F_{t+1} + C\tilde{Y}_t + u_{t+1}, \\ F_{t+1} = \mu + \varphi F_t + v_{t+1}, \end{cases} \quad [4.7]$$

where $\beta = (\beta_1, ..., \beta_n)'$ and $C = (c_{i,j})$.

Let us now compute the first-order and second-order moments conditional on the information on current and lagged excess returns $\underline{\tilde{Y}_t} = (\tilde{Y}_t, \tilde{Y}_{t-1}, ...)$. When the factor is static, i.e. $\varphi = 0$, we have $EF_t = \mu$ and the conditional expected excess returns are:

$$E_t(\tilde{Y}_{t+1}) = \mu\beta + C\tilde{Y}_t. \qquad [4.8]$$

The conditional variances and covariances between the excess returns are:

$$V_t(\tilde{y}_{i,t+1}) = \sigma^2 + \eta^2\beta_i^2, \quad Cov_t(\tilde{y}_{i,t+1}, \tilde{y}_{j,t+1}) = \eta^2\beta_i\beta_j, i \neq j.$$

These conditional variances and covariances can be written in a matrix:

$$V_t(\tilde{Y}_{t+1}) \equiv \Omega = \sigma^2 Id + \eta^2\beta\beta',$$

which is equal to the marginal variance-covariance matrix obtained in the static case.

Let us now consider the general framework with $\varphi \neq 0$. We can distinguish the first-order and second-order conditional moments depending on the information set available at time t. Generally speaking, the information set can be either $I_t = (\underline{\tilde{Y}_t}, \underline{F_t})$ or $J_t = (\underline{\tilde{Y}_t})$. In financial applications, it is usually assumed that the investor has more information than the econometrician[1]. Thus, I_t (respectively, J_t) is often interpreted as the information of the investor (respectively, of the econometrician). Let us first derive the first-order and second-order moments for the investor, that is with observable factor. We obtain:

$$E(\tilde{Y}_{t+1}|\underline{\tilde{Y}_t}, \underline{F_t}) = (\mu + \varphi F_t)\beta + C\tilde{Y}_t,$$

which can differ from the unconditional expectation, and:

$$V(\tilde{Y}_{t+1}|\underline{\tilde{Y}_t}, \underline{F_t}) \equiv \Omega = \sigma^2 Id + \eta^2\beta\beta',$$

1 Another information set might be introduced for a highly informed investor by considering the set $(\underline{\tilde{Y}_t}, \underline{F_{t+1}})$. We prefer the choice of I_t in order to compare static and dynamic portfolio managements.

which is still the marginal variance-covariance matrix obtained in the static case. Closed form expressions of the moments conditional on $\underline{\tilde{Y}_t}$ only, which are the moments for the econometrician, are not available. However, they can be computed numerically by the Kalman's filter (see section 3.7.3).

4.1.2. *Mean-variance portfolio management*

Let us now explain how a portfolio could be managed, that is regularly updated, by balancing between expected return and risk.

4.1.2.1. *Portfolio excess return*

Let us assume that a portfolio manager invests in a risk-free asset with a $y_{0,t}$ risk-free rate at date t and in n risky assets with returns Y_t on period $(t, t+1)$. At date t, he/she allocates the total budget W_t between the $n+1$ assets: $W_t = a_{0,t} + \sum_{i=1}^{n} a_{i,t}$, where $a_{0,t}$ (respectively, $a_{i,t}$, $i = 1, ..., n$) is the value invested in the risk-free asset (respectively, risky asset i, $i = 1, ..., n$). At date $t+1$, the portfolio value becomes:

$$\begin{aligned} W_{t+1} &= a_{0,t}(1 + y_{0,t}) + \sum_{i=1}^{n} a_{i,t}(1 + y_{i,t+1}) \\ &= W_t(1 + y_{0,t}) + \sum_{i=1}^{n} a_{i,t}(y_{i,t+1} - y_{0,t}). \end{aligned}$$

We deduce the portfolio return as:

$$y_{p,t+1} = (W_{t+1} - W_t)/W_t = y_{0,t} + \sum_{i=1}^{n} \delta_{i,t}(y_{i,t+1} - y_{0,t}),$$

where $\delta_{i,t} = a_{i,t}/W_t$, $i = 1, ..., n$, denote the fractions of the budget invested in each risky asset. These fractions do not sum up to 1 in general, since a part of the budget is invested in the risk-free asset. Moreover, some $\delta_{i,t}$'s can be negative, since it is possible to borrow in risky assets (no short sell restriction). If we denote by $\tilde{y}_{p,t+1} = y_{p,t+1} - y_{0,t}$ (respectively, $\tilde{y}_{i,t+1} = y_{i,t+1} - y_{0,t}$) the excess returns of the portfolio (respectively, risky asset i), we obtain:

$$\tilde{y}_{p,t+1} = y_{p,t+1} - y_{0,t} = \sum_{i=1}^{n} \delta_{i,t}\tilde{y}_{i,t+1} = \delta_t'\tilde{Y}_{t+1},$$

where $\delta_t = (\delta_{1,t}, \ldots, \delta_{n,t})'$.

4.1.2.2. *Mean-variance efficient portfolio*

Thus, the first-order and second-order conditional moments of the portfolio excess returns are:

$$E_t(\tilde{y}_{p,t+1}) = E_t(\delta_t'\tilde{Y}_{t+1}) = \delta_t' \, E_t(\tilde{Y}_{t+1}),$$
$$V_t(\tilde{y}_{p,t+1}) = V_t(\delta_t'\tilde{Y}_{t+1}) = \delta_t' \, V_t(\tilde{Y}_{t+1}) \, \delta_t.$$

At date t, the mean-variance optimization problem gives the efficient allocation δ_t in risky assets as the solution of:

$$\max_{\delta_t} \delta_t' \, E_t(\tilde{Y}_{t+1}) - A \, \delta_t' \, V_t(\tilde{Y}_{t+1}) \, \delta_t,$$

where $A > 0$. The objective function is an increasing function of the expected portfolio return and a decreasing function of the portfolio risk. The balance between these components depends on a parameter A, which can be interpreted as the (absolute) risk aversion of the investor.

The objective function is a concave function of the portfolio allocation. Thus, the first-order condition is necessary and sufficient to get the optimal allocation. The associated first-order condition is:

$$E_t(\tilde{Y}_{t+1}) - 2AV_t(\tilde{Y}_{t+1})\delta_t = 0,$$

and the vector of mean-variance efficient allocations in the n risky assets is proportional to [MAR 52]:

$$\delta_t \propto V_t(\tilde{Y}_{t+1})^{-1} E_t(\tilde{Y}_{t+1}),$$

with proportionality coefficient equal to half the inverse of the absolute risk aversion coefficient $1/A$. The larger the risk aversion coefficient, the smaller the fractions invested in risky assets will be.

Let us now discuss the form of the efficient allocation, when the asset excess returns satisfy either the static or the dynamic model of section 4.1.1.

1) *Static model ($\varphi = 0$, $C = 0$)*

Due to the static assumption, the conditional and unconditional moments coincide, and the mean-variance efficient allocation is time independent, given by:

$$\begin{aligned}\delta &= V(\tilde{Y}_{t+1})^{-1} E(\tilde{Y}_{t+1}) = \Omega^{-1}\mu\beta, \\ &= \left[\frac{1}{\sigma^2}\left(Id - \frac{\beta\beta'}{\beta'\beta}\right) + \frac{1}{\sigma^2 + \eta^2\beta'\beta}\frac{\beta\beta'}{\beta'\beta}\right]\mu\beta, \\ &= \frac{\mu\beta}{\sigma^2 + \eta^2\beta'\beta}.\end{aligned}$$

The vector of efficient allocations in risky assets is proportional to the vector of beta's.

The associated Sharpe performance, which is the marginal expected return adjusted for risk of the n risky assets [SHA 66], is:

$$\begin{aligned}S &= E(\tilde{Y}_{t+1})' V(\tilde{Y}_{t+1})^{-1} E(\tilde{Y}_{t+1}) \\ &= \mu^2\beta'\left[\frac{1}{\sigma^2}\left(Id - \frac{\beta\beta'}{\beta'\beta}\right) + \frac{1}{\sigma^2 + \beta'\beta\eta^2}\frac{\beta\beta'}{\beta'\beta}\right]\beta\end{aligned}$$

$$= \frac{\mu^2}{\sigma^2 + \beta'\beta\eta^2}\beta'\beta$$

$$= \frac{\mu^2\beta'\beta}{\sigma^2 + \eta^2\beta'\beta}.$$

The Sharpe performance depends on the beta's by means of $\beta'\beta$ only and is an increasing function of this quantity.

2) *Dynamic model*

Due to the dynamic assumption, the conditional and unconditional moments do not coincide and the efficient allocation becomes time-dependent. We obtain:

$$\delta_t = \frac{\mu\beta}{\sigma^2 + \eta^2\beta'\beta} + \frac{\varphi F_t\beta}{\sigma^2 + \eta^2\beta'\beta}$$
$$+ \left[\frac{1}{\sigma^2}\left(Id - \frac{\beta\beta'}{\beta'\beta}\right) + \frac{1}{\sigma^2 + \beta'\beta\eta^2}\frac{\beta\beta'}{\beta'\beta}\right] C\tilde{Y}_t.$$

The vector of efficient allocations is the sum of three terms. The first term is the optimal allocation in the static case, whereas the two other terms adjust the optimal allocation to include the fact that current excess returns depend on past factor value and past excess returns, respectively.

4.1.3. *Immunized portfolios and mimicking portfolios*

Let us now look for other types of portfolio management. First we consider portfolios which are the least sensitive to a risk factor. They are called factor neutral portfolios. We also say that they are hedged or immunized against risk factor. Second we are interested in portfolios with returns close to the factor values, which are (factor) mimicking portfolios. They are used, for instance, to construct new generations of index funds.

4.1.3.1. *Immunized portfolios*

Let us consider the general dynamic model with both common factor and contagion, and focus on the immunized portfolios. The excess return of a portfolio with γ allocation in the risky assets is:

$$\gamma'\tilde{Y}_{t+1} = \gamma'\beta F_{t+1} + \gamma' C\tilde{Y}_t + \gamma' u_{t+1}.$$

This quantity does not depend on factor values if and only if $\gamma'\beta = 0$, that is, if γ belongs to the vector space generated by the vector of betas. The immunized portfolios are such that:

$$\gamma'\tilde{Y}_{t+1} = \gamma' C\tilde{Y}_t + \gamma' u_{t+1}. \qquad [4.9]$$

In particular, by considering these directions, we also eliminate the effect on $\tilde{Y}_{t+1}$ of lagged values of $\tilde{Y}$ with a lag larger or equal to 2. This property can be used to directly identify these allocations from the autocovariance function of $\tilde{Y}_t$. Let us consider the linear regression of $\gamma'\tilde{Y}_{t+1}$ on both $\tilde{Y}_t, \tilde{Y}_{t-1}$:

$$\gamma'\tilde{Y}_{t+1} = C_1(\gamma)\tilde{Y}_t + C_2(\gamma)\tilde{Y}_{t-1} + \tilde{u}_{t+1}, \text{ say.} \qquad [4.10]$$

The theoretical regression coefficients are:

$$[C_1(\gamma), C_2(\gamma)] = Cov\left[\gamma'\tilde{Y}_{t+1}, \begin{pmatrix} \tilde{Y}_t \\ \tilde{Y}_{t-1} \end{pmatrix}\right] \left[V\begin{pmatrix} \tilde{Y}_t \\ \tilde{Y}_{t-1} \end{pmatrix}\right]^{-1}$$

$$= \gamma' \left[\Gamma_Y(1), \Gamma_Y(2)\right] \begin{bmatrix} \Gamma_Y(0) & \Gamma_Y(1) \\ \Gamma_Y(1)' & \Gamma_Y(0) \end{bmatrix}^{-1},$$

where $\Gamma_Y(h) = Cov\left(\tilde{Y}_{t+1}, \tilde{Y}_{t-h}\right)$. We deduce that:

$$C_2(\gamma) \equiv \gamma' C_2,$$

where $C_2 = [\Gamma_Y(1) - \Gamma_Y(2)\Gamma_Y(0)^{-1}\Gamma'_Y(1)][\Gamma_Y(0) - \Gamma_Y(1) \Gamma_Y(0)^{-1}\Gamma_Y(1)']^{-1}$ by inverting by blocks. The matrix C_2 is a multivariate partial autocovariance of order 2 (see [RAM 74]).

By comparing equations [4.9] and [4.10], we note that the allocations of the immunized portfolios are such that:

$$C_2(\gamma) = 0 \Leftrightarrow \gamma' C_2 = 0 \Leftrightarrow C_2'\gamma = 0.$$

Thus, under the identification restriction $Rank\ C_2 = 1$, the immunized allocations are the vectors belonging to the null space of C_2', whose dimension is $n - 1$.

In practice, the matrix C_2 can be estimated by substituting in the expression of C_2 the theoretical autocovariances by their sample counterparts. Due to the estimation errors, the estimated matrix $\hat{C}_2$ is of full rank, but the direction generating the space orthogonal to the set of immunized allocations can be consistently estimated by the eigenvector associated with the largest eigenvalue of $\hat{C}_2\hat{C}_2'$.

In the static case $\varphi = 0$, $C = 0$, the autocovariances $\Gamma_Y(h) = 0$, $h > 0$. Thus, when $C_2 = 0$, the rank condition is not satisfied and the above estimation method does not apply.

4.1.3.2. *Mimicking portfolio*

Let us now look for portfolio allocation such that the evolution of the portfolio return mimics the evolution of the factor. For expository purposes, we assume $\mu = 0$, and focus on the allocation in risky assets only. This mimicking allocation is defined as the solution of the optimization problems:

$$\begin{aligned} \gamma_t^* &= \arg\min_\gamma E_t\left(\gamma' \tilde{Y}_{t+1} - F_{t+1}\right)^2 \\ &= \arg\min_\gamma V_t\left(\gamma' \tilde{Y}_{t+1} - F_{t+1}\right), \end{aligned} \qquad [4.11]$$

where E_t and V_t are the expectation and variance conditional on the past values of the observable returns $\underline{Y_t}$. The solution of the problem is the associated theoretical regression coefficient of F_{t+1} on $\tilde{Y}_{t+1}$:

$$\gamma_t^* = \left[V_t\left(\tilde{Y}_{t+1}\right)\right]^{-1} Cov_t\left(\tilde{Y}_{t+1}, F_{t+1}\right), \qquad [4.12]$$

and the associated portfolio return is:

$$\tilde{y}_{p,t+1} = Cov_t\left(F_{t+1}, \tilde{Y}_{t+1}\right)\left[V_t\left(\tilde{Y}_{t+1}\right)\right]^{-1}\tilde{Y}_{t+1}. \qquad [4.13]$$

In general, the factor mimicking allocation is a complicated function of past returns, and can only be computed numerically via the Kalman's Filter. However, this allocation becomes simple in a model without contagion, that is when $C = 0$. In the single factor case, we obtain:

$$Cov_t\left(F_{t+1}, \tilde{Y}_{t+1}\right) = Cov_t\left(F_{t+1}, \beta F_{t+1}\right) = V_t\left(F_{t+1}\right)\beta',$$
$$V_t\left(F_{t+1}\right) = V_t\left(\beta F_t + u_{t+1}\right) = V_t\left(F_t\right)\beta\beta' + \sigma^2 Id.$$

We deduce:

$$\gamma_t^* = \left(V_t\left(F_{t+1}\right)\beta\beta' + \sigma^2 Id\right)^{-1}\beta V_t\left(F_{t+1}\right). \qquad [4.14]$$

4.2. Contagion among banks

Linear dynamic models with common factor and contagion can be used as a first step to analyze the financial stability of a network of banks. This is usually done by means of stress tests, which consider the dynamic consequences of a shock (or stress). This shock can be specific, applied to an idiosyncratic error $u_{i,t}$, or be systematic, applied to the factor innovation. The dynamic models of this section allow us to disentangle the direct effects of the shock from the contagion effects by means of the contagion matrix C. They also allow us to understand the significant contagion channels and to discuss the associated contagion network.

As an illustration, we can implement the approach to the banking sector, with two types of information, either by considering the balance sheets of the firms, and, typically, their accounting equity value defined as the difference between their assets and liabilities in the firm's book, or by analyzing their market values when they are quoted on a stock exchange (see [DAR 14a]).

4.2.1. *Mixed frequencies*

Let us consider a perimeter of n banks, and stack the variables of interest, which can be either the rate of change in the book value, or in the market capitalization on period $(t-1,t)$, in a n-dimensional vector Y_t. These variables are preliminary demeaned and satisfy the dynamic factor model [3.1], which is[2]:

$$Y_t = BF_t + CY_{t-1} + u_t, \text{ with } F_t = AF_{t-1} + v_t, \qquad [4.15]$$

where the error terms u_t, v_t are zero-mean, serially independent, with second-order moments [3.2]. F_t gives the values at date t of K unobservable factors.

The variables of interest, either the rate of changes on the value of the firm, or the rate of changes on the capitalization, are dynamically dependent through the effects of the common exogenous factors, and through the effects of their lagged values. In practice, the underlying factors can include macroeconomic factors as well as unobserved effects of the other financial institutions, which are outside the perimeter of interest.

2 For expository purposes, we introduce no intercept in the equation. Indeed, whenever the process (Y_t) is stationary, this intercept can be set to zero by considering the demeaned series, $Y_t - \bar{Y}$, and, after this transformation, we can assume $E(F_t) = 0$.

The indicators of banks' financial vulnerabilities are not available at the same frequency. The indicators based on accounting data are accessible at a quarterly frequency, whereas the market-based indicators are available at a daily frequency. To facilitate the comparison with the analysis on market data, we consider the daily model [4.15] for the returns on book values. Due to the state space representation (see section 3.2.2), the model specification remains unchanged at a lower frequency:

$$Y_t = D_h F_t + C^h Y_{t-h} + u_{t,h}, F_t = A^h F_{t-h} + v_{t,h},$$

say, with $h = 60$ opening days for quarterly data frequency, and $D_h = CD_{h-1} + BA^h$ and $D_0 = 0$ (see section 3.7.4). This time coherency property of the model allows for the comparison of two series observed at different frequencies.

4.2.2. *Estimation results*

Darolles *et al.* [DAR 14a] focus on of the biggest banks in the eurozone, namely: Banco Santander SA, BNP Paribas SA, Commerzbank AG, Crédit Agricole SA, Deutsche Bank AG, Intesa San Paolo SPA, Société Générale SA and Unicredit SPA. Thus, Germany, France, Italy and Spain are countries represented in the sample, which allows us to investigate both intra- and international contagion among banks in the zone.

Data on institutions' market capitalization come from Bloomberg, while data on the banks' equity value are obtained from Bankscope. The equity value mixes mark-to-market valuation, especially for the asset component of the balance sheet, and contractual valuation for a part of the debt. Thus, we should expect a significant dynamic dependence between market capitalization and equity due to this accounting practice. The sample starts in January 2,

2007, and ends on December 20, 2012. We assume a single factor for expository purposes.

We provide in Table 4.1 the estimated contagion matrix C. We observe a total number of significant connections equal to 42 for the capitalization data and 50 for the accounting data, to be compared with 64 possible connections. Thus, this system of banks is highly interconnected. The different banks do not play the same role. For instance, the return on Commerzbank's stocks depends neither on its lagged return, nor on the returns of the other banks once the effect of the common factor has been taken into account. On the other hand, the Commerzbank has an effect on all the other banks. Another extreme example is Société Générale, which is affected by all the other banks for the two different series, and affects all the other banks, except the Commerzbank, for the changes in book values only.

	Santander	BNPP	Commerz	CASA	Deutsche	Intesa	SocGen	Unicredit
Santander	**−0.569**	**−0.165**	**0.081**	**0.250**	**0.437**	0.065	0.045	−0.111
	−0.042	**0.336**	0.110	**−0.193**	−0.160	**0.300**	**0.113**	**0.642**
BNPP	**0.527**	**−0.166**	**0.381**	**0.146**	**−0.902**	0.025	0.037	**0.274**
	−0.172	**0.168**	**0.234**	**−0.267**	**−0.211**	**0.251**	**0.391**	**0.510**
Commerz	0.092	0.059	−0.046	0.004	0.050	0.051	0.018	−0.083
	0.225	**0.101**	0.063	**0.525**	**0.299**	−0.012	−0.011	**−0.185**
CASA	**0.213**	**0.302**	**−0.094**	**−0.308**	−0.088	0.045	0.099	**−0.119**
	0.027	**0.126**	**0.104**	**−0.326**	0.115	**0.101**	**0.098**	0.004
Deutsche	−0.088	0.008	**0.633**	**−0.731**	**0.471**	**−0.280**	**−0.242**	**0.646**
	0.337	−0.094	**−0.183**	**0.190**	**0.158**	**0.179**	**0.179**	**0.299**
Intesa	**0.172**	**0.247**	**0.281**	**0.233**	**0.217**	**0.138**	**0.253**	**0.324**
	−0.116	**−0.104**	**−0.088**	**−0.456**	0.194	**−0.215**	**−0.277**	**−0.497**
SocGen	**−0.336**	**−0.245**	**−0.267**	**0.526**	**0.292**	**0.142**	**0.202**	**−0.195**
	−0.303	**−0.135**	**−0.299**	**0.454**	**−0.204**	**−0.365**	**−0.289**	**−0.118**
Unicredit	0.109	**0.199**	−0.038	**0.144**	**−0.139**	0.105	0.069	**−0.186**
	−0.024	**−0.127**	**−0.227**	**0.191**	**0.386**	0.056	**0.213**	**−0.139**

Table 4.1. *Contagion matrix obtained from banks' market capitalization (first row in each cell), or accounting data (second row), at quarterly frequency. Significant parameters (at 5%) are in bold (from [DAR 14a])*

The paths of the estimated latent factors are displayed in Figure 4.1 for the two datasets. The two dynamic frailties feature negative autocorrelations equal to −0.22 and −0.65 for the capitalization and the book series, respectively. They share a certain degree of commonality, with a cross-correlation equal to 0.45. Nevertheless, we see in Figure 4.1 that the two factor paths have rather different nonlinear features. For instance, rather large volatilities are observed for book values between 2008 and 2009, but after 2011, they are difficult to detect in the evolution of the factor for stock returns.

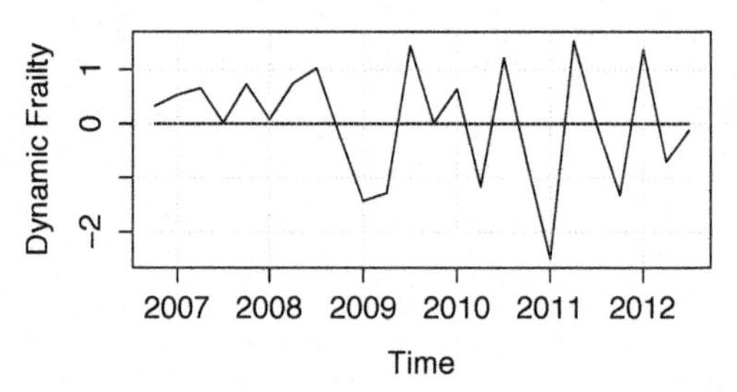

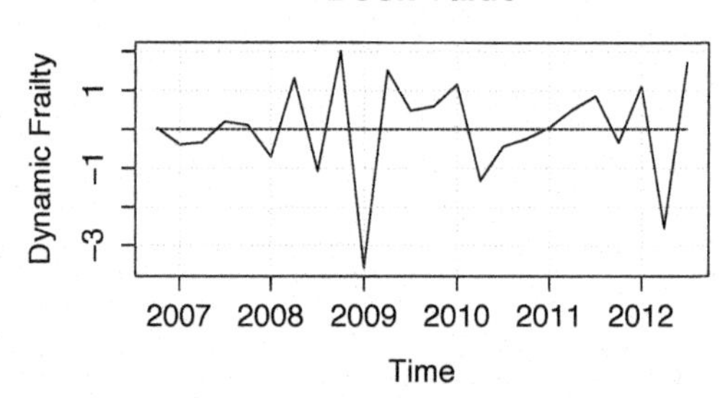

Figure 4.1. *Estimated dynamic frailties obtained from banks' market capitalization (top), book values (bottom), at quarterly frequency (from [DAR 14])*

The estimated factor betas are given in Table 4.2. As expected from financial theory, the capitalization common factor affects all stock returns with the largest influence on the Commerzbank.

	Santander	BNPP	Commerz	CASA	Deutsche	Intesa	SocGen	Unicredit
Capitalization	0.1530**	0.216*	0.335**	0.216**	0.190**	0.160**	0.204**	0.208**
	(0.049)	(0.116)	(0.106)	(0.075)	(0.064)	(0.071)	(0.051)	(0.071)
Book value	0.029	0.065	0.084*	0.091	0.026	0.042	0.037	0.028
	(0.031)	(0.044)	(0.046)	(0.090)	(0.038)	(0.051)	(0.052)	(0.031)

Table 4.2. *Factor betas obtained from banks' market capitalization, or accounting data (* significant at 10%, ** significant at 5%) (standard errors in parentheses) (from [DAR 14a])*

However, the interpretation of the frailty as a common factor is no longer valid when we consider the analysis based on book values, since in this case only one beta coefficient, i.e. the beta for Commerzbank, is significant. The interpretation of the significant beta is now different. If all betas were equal to 0, the dynamic model would be a VAR(1) model, with short memory features. The presence of a significant beta introduces longer memory, and this long memory is channeled through the Commerzbank.

4.2.3. *Impulse response*

Let us now analyze the effects of a systematic shock, which is a shock on the common factor, and decompose these effects to highlight the role of contagion. We provide in Figure 4.2 the impulse response functions for the capitalization of the banks Intesa and BNPP. The direct effect of the initial shock (striped area) is comparable for the two banks, and concentrated on the two first dates. However, contagion phenomena (full area) have different effects among banks. For Intesa, the contagion amplifies the initial negative direct impact, both in size and duration. For BNPP, the initial shock is diffused and increases the stock volatility during the next 10 quarters.

We compare in Figure 4.3 the direct and indirect impacts for all the banks in the network. On the top panel, the total effect of the same initial shock is different from one bank to another. If the initial impact on bank capitalization is negative for all banks, we observe a huge heterogeneity within the network for the next dates. To understand the source of this observed heterogeneity, we plot in the two following panels the direct and indirect components of the impulse responses. The direct effect is only driven by the loading factors and the factor persistence. The mean reverting behavior observed on the common factor explains the positive effect of the exogenous shock for $h = 1$. This is a short-term effect and vanishes after three quarters.

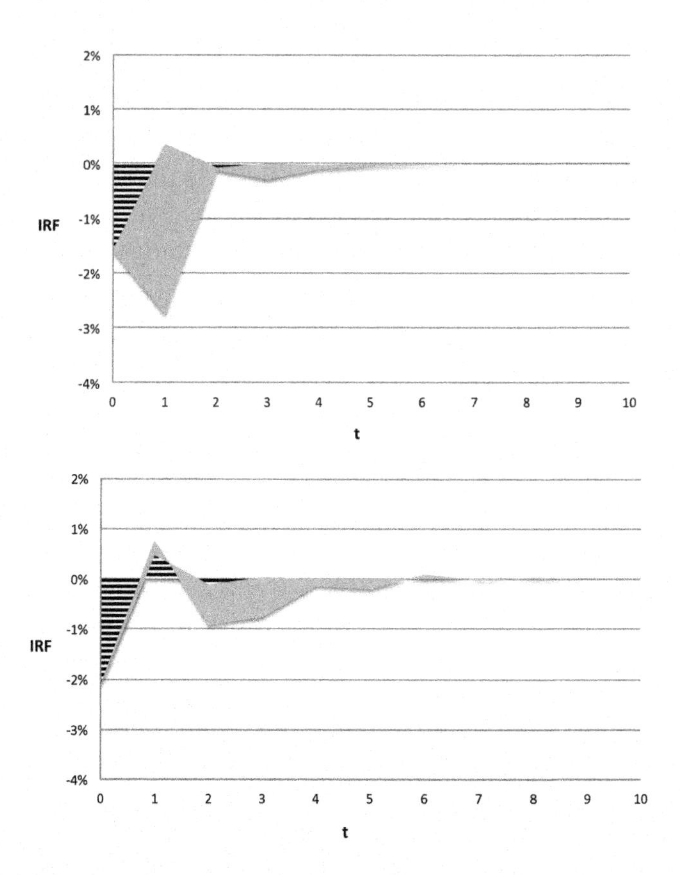

Figure 4.2. *Impulse response functions – Intesa (top) versus BNPP (bottom) for capitalization. This figure displays direct (striped area) and indirect (full area) effects of an initial shock (from [DAR 14])*

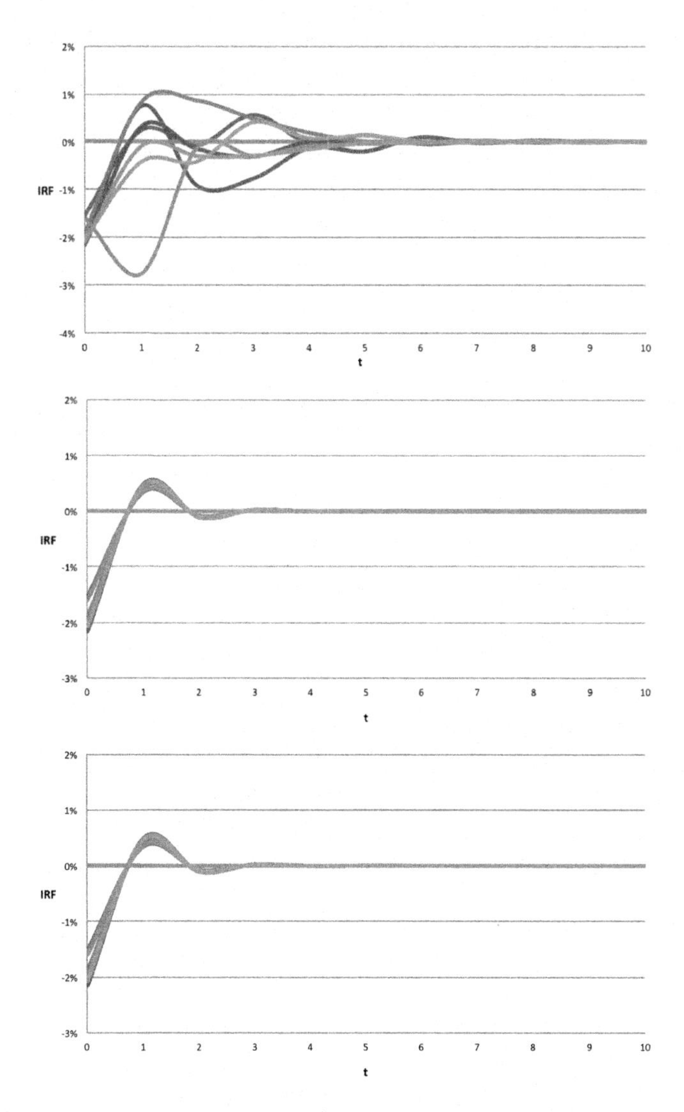

Figure 4.3. *Impulse response functions. This figure displays the total (top), direct (middle) and indirect (bottom) effects of an initiation shock on banks capitalization – Santander (blue), BNPP (red), Commerzbank (green), CASA (violet), Deutsche (turquoise), Intensa (orange), SocGen (light blue) and Unicredit (brown) (from [DAR 14]). For a color version of this figure, see www.iste.co.uk/darolles/contagion.zip*

The indirect effect is more complex. It arises from the contagion matrix that creates a complicated dynamic for the diffusion of the initial shock. For Intesa, the contagion effect is only negative and concentrated on a short horizon. For BNPP and Santander, we observed oscillations that increase the idiosyncratic volatility of the two stocks. These three banks are exposed to contagion and appear vulnerable in this respect. On the contrary, for Unicredit and SocGen, the contagion effect stays close to zero at all maturities. These two banks have low vulnerability to contagion even if they have a significant exposure to the initial adverse shock. We also observe that for all banks direct and indirect effects do not have the same maturity.

The direct persistence of the initial shock is high for book data (see Figure 4.4), due the highest factor persistence. Thus, this effect is significant during a longer period that is equivalent of the period we observe a significant indirect effect. However, if the direct effect is equivalent for all banks in the network, the indirect effect is heterogeneous and generates dispersion between banks. Commerzbank and Intesa are the two most vulnerable banks, while Unicredit and SocGen are the least vulnerable banks.

4.3. Chapter 4 highlights

The dynamic linear factor model with both frailty and contagion can be used for portfolio management of basic assets, such as stocks, bonds and currencies. We have described different types of management: mean-variance efficient portfolios, portfolios immunized against the factor variation and factor mimicking portfolios. However, these approaches have to be modified for portfolios including derivatives such as European options, swaps, futures and credit default swaps (CDS). Indeed, these derivatives have nonlinear payoffs and their analysis requires nonlinear dynamic models (see Chapter 5).

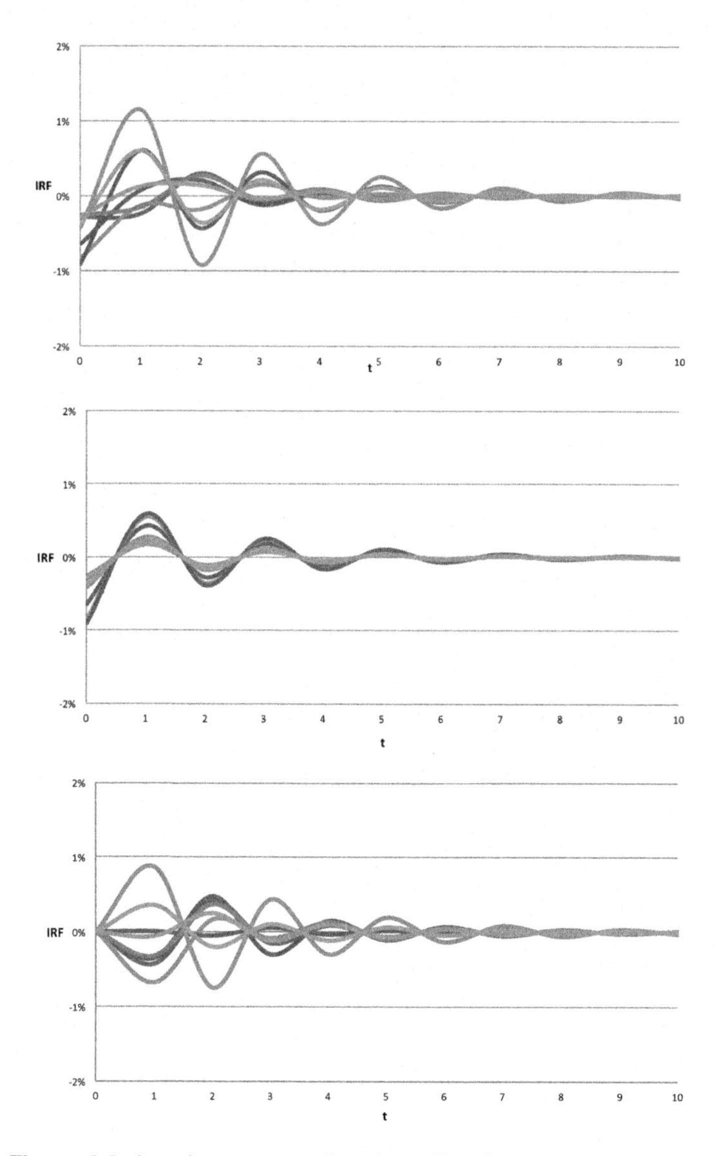

Figure 4.4. *Impulse response functions. This figure displays the total (top), direct (middle) and indirect (bottom) effects of an initiation shock on banks book value – Santander (blue), BNPP (red), Commerzbank (green), CASA (violet), Deutsche (turquoise), Intensa (orange), SocGen (light blue) and Unicredit (brown) (from [DAR 14]). For a color version of this figure, see www.iste.co.uk/darolles/contagion.zip*

We have also illustrated the use of linear dynamic models to show how to disentangle the direct and indirect (contagion) effects of exogenous shocks on the financial situations of a set of banks. This modeling has been applied to a set of eight banks and their financial vulnerability measured by the change in either their market capitalization or their book values. Even if we might expect similar results for the two different measures, the analysis shows that the interconnections revealed from market data and accounting data, respectively, are significantly different.

In practice, the European banking system contains much more than eight banks and the extension of the methodology will require methods for large-scale factor models based on sparse estimation or the Lasso approach, for instance (see [BAR 13] for such an approach in a model without frailty).

4.4. Appendices

4.4.1. *Decomposition of the variance-covariance matrix*

Let us recall that an orthogonal projector is a square matrix P, which is symmetric $P' = P$ and idempotent $P^2 = P$. The variance-covariance matrix $\Omega = \sigma^2 Id + \eta^2 \beta\beta'$ (see equation [4.4]) is equivalently written as: $\Omega = \sigma^2(Id - P) + \lambda^2 P$, where $\lambda^2 = \sigma^2 + \eta^2 \beta'\beta$ and $P = \beta\beta'/\beta'\beta$.

We check that $P' = P$ and $P^2 = \frac{\beta\beta'}{\beta'\beta}\frac{\beta\beta'}{\beta'\beta} = \frac{\beta\beta'}{\beta'\beta} = P$. P is the orthogonal projector on the vector space generated by the vector of beta coefficients, whereas $Id - P$ is the projector on the vector space orthogonal to the vector of betas.

This decomposition can be used directly to find the inverse of the matrix Ω. Indeed, let us consider the matrix given in equation [4.6]:

$$A = \frac{1}{\sigma^2}(Id - P) + \frac{1}{\lambda^2}P.$$

We have:

$$
\begin{aligned}
A\Omega &= \left(\frac{1}{\sigma^2}(Id - P) + \frac{1}{\lambda^2}P\right)\left(\sigma^2(Id - P) + \lambda^2 P\right) \\
&= (Id - P)^2 + P^2 \\
&\qquad \text{(since } P(Id - P) - P - P^2 = 0, \text{ by idempotence)} \\
&= Id - 2P + 2P^2 \\
&= Id \text{ (by idempotence).}
\end{aligned}
$$

We deduce that $A = \Omega^{-1}$.

5

Common Frailty and Contagion in Nonlinear Dynamic Models

The linear dynamic models are rather simple, but cannot take into account the nonlinear features existing in risk analysis. These nonlinearities are due to the derivatives traded on the market: for instance, a European call option has a nonlinear payoff $\max(S_T, K)$, where S_T is the value of the underlying asset at maturity T and K is the strike of the option. Thus, this payoff is a piecewise linear function of the stochastic variable S_T. They can also result from qualitative events such as defaults, prepayments, interventions of a Central Bank, generally represented by $0 - 1$ indicators. They also have to be introduced to capture the risk premium, i.e. the effect of the volatility of a return (a squared return) on the expected return.

In this chapter, we extend to nonlinear dynamic models the notions of common factor (called dynamic frailty) and contagion. We apply this extension to the analysis of stochastic volatility models.

5.1. Specifications

The nonlinear dynamics can be specified in equivalent alternative ways, based on either a joint transition density, or nonlinear regressions, or conditional Laplace transforms. We describe these alternative modeling strategies in this section. The choice between the specifications depends on the estimation approach, Bayesian or classical, based on a maximum likelihood or a moment method, and on the purpose of the analysis, that can be nonlinear filtering of the factors, or nonlinear forecasting of the variables of interest: returns, volatilities.

All the specifications share the same main features as the linear dynamic models. They include lagged observed values, systematic and idiosyncratic factors, and an exogeneity condition. Moreover, the dependence in lagged values has to make appear a contagion matrix to be estimated.

For expository purpose, but also to avoid the curse of dimensionality, we consider Markov dynamics, that is, we assume that the distribution of the pair (Y_t, F_t) given the past $\underline{Y_{t-1}} = (Y_{t-1}, Y_{t-2}, ...)$, $\underline{F_{t-1}} = (F_{t-1}, F_{t-2}, ...)$, depends on the past through the most recent values (Y_{t-1}, F_{t-1}), only. Under the Markov assumption, the dynamics is characterized by the transition distribution, that is, by the joint conditional distribution of (Y_t, F_t) given[1] (Y_{t-1}, F_{t-1}).

---

1 Let us consider a Markov process (X_t) with continuous transition density, say $l(x_t|x_{t-1})$. Under stability conditions on the transition density, the process (X_t) can be chosen stationary with a stationary density $l_0(x)$ solution of the Kolmogorov integral equation: $l_0(x) = \int l(x|\tilde{x}) l_0(\tilde{x}) d\tilde{x}$. Thus, if the solution of this integral equation is unique, the stationary distribution, as well as the complete distribution of the process, are characterized by the transition density.

5.1.1. *Transition-based specification*

By the Bayes formula, the joint transition can always be decomposed as:

$$l\left(y_t, f_t | y_{t-1}, f_{t-1}\right) = l\left(y_t | y_{t-1}, f_t, f_{t-1}\right) l\left(f_t | y_{t-1}, f_{t-1}\right). \quad [5.1]$$

Let us now parametrize this joint transition. The parameter θ includes parameters C, β in $l\left(y_t | y_{t-1}, f_t, f_{t-1}\right)$ and parameters α in $l\left(f_t | y_{t-1}, f_{t-1}\right)$, where C has the interpretation of a contagion matrix. This parametrization is summarized in the following assumptions:

ASSUMPTION A.1 (transition equation).–
$l(f_t | y_{t-1}, f_{t-1}; \theta) = l(f_t | f_{t-1}; \alpha)$.

ASSUMPTION A.2 (measurement equation).–
$l(y_t | y_{t-1}, f_t, f_{t-1}; \theta) = l(y_t | f_t, C y_{t-1}; \beta)$.

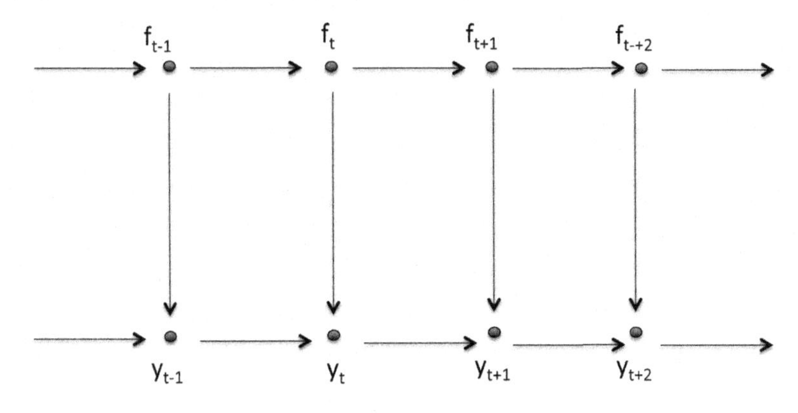

Figure 5.1. *Causal scheme*

Thus, we get the causal scheme displayed in Figure 5.1. The common factor F_t has an exogenous dynamics, since its conditional distribution depends on the past by means of F_{t-1} only. Moreover, given the current value F_t of the exogenous factor, the effect of Y_{t-1} on Y_t passes through the linear transformation CY_{t-1}, where C can be interpreted as a contagion matrix.

5.1.2. *Regression-based specification*

The proposition below provides the equivalence between a specification based on the joint transition density and a specification based on regressions. Moreover, the proof of this proposition explains how to derive these nonlinear regressions in practice (see section 5.5.1).

PROPOSITION 5.1.– Let us assume that the variables $(Y_t, Y_{t-1}, F_t, F_{t-1})$ have a joint continuous distribution. The assumptions A.1–A.2 can be equivalently written as:

ASSUMPTION A.1 (transition equation).– There exists a standard Gaussian noise $v_t \sim IIN(0, Id_K)$ independent of $\underline{F_{t-1}}, \underline{Y_{t-1}}$ and a function b such that:

$$F_t = b\left(F_{t-1}, v_t; \alpha\right), \qquad [5.2]$$

where $b\left(F_{t-1}, ; \alpha\right)$ is one-to-one in v.

ASSUMPTION A.2 (measurement equation).– There exists a standard Gaussian noise $u_t \sim IIN(0, Id_n)$ independent of $\underline{F_t}, \underline{Y_{t-1}}$ and a function a such that:

$$Y_t = a\left(F_t, CY_{t-1}, u_t; \beta\right), \qquad [5.3]$$

where $a\left(F_t, CY_{t-1}, ; \beta\right)$ is one-to-one in u.

v_t is called the (causal) nonlinear innovation of factor F_t (see [ROS 00]).

The differences between the nonlinear regressions in proposition 5.1 and the linear dynamic model of Chapter 3 are twofold. Firstly, nonlinear regression [5.2] allows for cross-effects between systematic and idiosyncratic factors, between systematic factors and contagion, whenever regression [5.2] does not admit an additive decomposition such as:

$$Y \equiv a_1(F_{t-1};\alpha) + a_2(CY_{t-1};\alpha) + a_3(u_t;\alpha), \text{ say.} \quad [5.4]$$

Secondly, even if the additive decomposition [5.4] were assumed, functions a_1, a_2, a_3 would not be linear in general. For instance, if a_1, a_2 are linear, but a_3 is not linear, we get a linear regression model with a non-Gaussian error term:

$$Y = BF_t + CY_{t-1} + a_3(u_t;\alpha), \text{ say.} \quad [5.5]$$

The regression-based specification [5.2]–[5.3] is especially useful for simulating paths of process (Y_t, F_t), for fixed parameter values. The simulated steps are the following ones:

Step 1: Draw independently values $u_t^s, v_t^s, t = -100, \ldots, T$ in the standard Gaussian distributions of dimensions n and K, respectively (that is, select different seed numbers for u and for v, and apply a Gaussian generator).

Step 2: Compute recursively:

$$F_t^s = b\left(F_{t-1}^s, v_t^s; \beta\right), \ t = -100, \ldots, T,$$
$$Y_t^s = a\left(F_t^s, CY_{t-1}^s, u_t^s; \alpha\right), t = -100, \ldots, T,$$

with starting values $Y_{-101}^s = 0, F_{-101}^s = 0$.

Step 3: Keep the simulated paths (Y_t^s, F_t^s), $t = 1, \ldots, T$.

This approach is valid whenever process (Y_t, F_t) is strictly stationary. The simulation starts at $t = -100$, to ensure that Y_0, F_0 are almost drawn in the stationary distribution of (Y_t, F_t).

5.1.3. *Laplace transform-based specification*

The distribution of any multidimensional variable, continuous or discrete, can be characterized by its Laplace transform. Let us denote Y this variable, the Laplace transform is the function:

$$\Psi_Y(z) = E\left[\exp(z'Y_t)\right], \qquad [5.6]$$

defined for any complex multidimensional argument z such that the expectation in the right hand side exists. When z is restricted to real vectors (respectively, vectors with pure imaginary components), the Laplace transform is called moment generating function (respectively, characteristic function).

In terms of (conditional) Laplace transform, assumptions A.1 and A.2 become:

ASSUMPTION A.1 (transition equation).–

$$E\left[\exp(w'F_t)\middle|\, Y_{t-1}, F_{t-1}; \theta\right] = E\left[\exp(w'F_t)\middle|\, F_{t-1}; \beta\right], \forall w.$$

ASSUMPTION A.2 (measurement equation).–

$$E\left[\exp(z'Y_t)\middle|\, \underline{Y_{t-1}}, F_t, F_{t-1}; \theta\right] = E\left[\exp(z'Y_t)\middle|\, F_t, CY_{t-1}; \alpha\right], \forall z.$$

The specification by means of the conditional Laplace transforms is especially appropriate to model the dynamics of term structure of interest rates, and of term structure of defaults (see e.g. [DUF 03, GOU 07]).

5.1.4. *Gaussian linear dynamic model*

Let us now explicate the above alternative specifications for the Gaussian linear dynamic model of Chapter 3.

i) The regression-based specification is simply the state-space representation of (Y_t, F_t):

$$\text{transition equation: } F_t = AF_{t-1} + \Omega^{1/2} v_t, v_t \sim N(0, Id_K),$$

$$\text{measurement equation: } Y_t = BF_t + CY_{t-1} + \Sigma^{1/2} u_t,$$
$$u_t \sim N(0, Id_n).$$

ii) The specification by the transition densities is:

$$l(f_t|f_{t-1}) = \frac{1}{(2\pi)^{K/2}\sqrt{\det \Omega}} \times \exp\left[-\frac{1}{2}(f_t - Af_{t-1})'\Omega^{-1}(f_t - Af_{t-1})\right],$$

$$l(y_t|f_t, y_{t-1}) = \frac{1}{(2\pi)^{n/2}\sqrt{\det \Sigma}} \times \exp\left[-\frac{1}{2}(y_t - Bf_t - Cy_{t-1})'\Sigma^{-1} \times (y_t - Bf_t - Cy_{t-1})\right].$$

iii) Finally, the associated conditional Laplace transforms are:

transition equation:

$$E\left[\exp(w'F_t)|f_{t-1}\right] = \exp\left[w'Af_{t-1} + \frac{w'\Omega w}{2}\right],$$

measurement equation:

$$E\left[\exp(z'Y_t)|f_t, y_{t-1}\right] = \exp\left[z'(Bf_t + Cy_{t-1}) + \frac{z'\Sigma z}{2}\right].$$

The logarithms of the conditional Laplace transforms are affine functions of the conditioning variables. When such an affine condition is satisfied, the nonlinear dynamic model is called affine model or compound autoregressive model (CAR) (see e.g. [DUF 03, DAR 06]). Such affine specifications are appropriate for the analysis of term structures of risk, contagion, rates, etc. Indeed under an affine specification, closed form term structures are frequently available.

5.2. Stochastic volatility model

In order to illustrate the modeling strategies in a nonlinear dynamic framework, let us consider a model for the joint analysis of returns allowing for stochastic volatility. Such a model is a multivariate extension of the one-dimensional stochastic volatility model introduced by Hull and White [HUL 87], Heston [HES 93] and Ball and Roma [BAL 94]. Y_t denotes the n-dimensional vector of returns, Σ_t denotes its (n, n) volatility–covolatility matrix. In such models, the elements of matrix Σ_t are the volatilities: $V_{t-1}(Y_{i,t}), i = 1, ..., n$ on the diagonal, and the covolatilities: $Cov_{t-1}(Y_{i,t}, Y_{j,t}), i \neq j$, out of the diagonal. They play the role of unobserved systematic factors. Thus, for a joint analysis of $n = 10$ stock returns, the number of such factors is equal to $n(n+1)/2 = 55$.

5.2.1. *The return dynamics given the volatility path*

This specification is:

$$Y_t = CY_{t-1} + \begin{bmatrix} Tr(B_1\Sigma_t) \\ \vdots \\ Tr(B_n\Sigma_t) \end{bmatrix} + \Sigma_t^{1/2} u_t, \qquad [5.7]$$

where (u_t) is a standard Gaussian noise independent of the volatility process (Σ_t); $C, B_1, ..., B_n$ are (n, n) matrices and Tr denotes the trace operator[2]. Specification [5.7] is the natural

2 The trace of a square matrix is the sum of its diagonal elements: $Tr(A) = \sum_{i=1}^{n} a_{i,i}$. In formula [5.7] appear traces of product of matrices: $Tr(BA)$. Such a trace is equal to: $Tr(BA) = \sum_{i=1}^{n}(BA)_{i,i} = \sum_{i=1}^{n}\sum_{j=1}^{n} b_{i,j}a_{j,i}$, and

extension of the basic Gaussian autoregressive models for returns, when the volatilities and covolatilities are not constant. A given equation of [5.7] says that the return of asset i depends on the lagged returns of all assets (the contagion effect), and of the current volatilities and cross volatilities:

$$Y_{i,t} = \sum_{j=1}^{n} c_{i,j} Y_{j,t-1} + \sum_{j=1}^{n} b^i_{j,j} V_{t-1}(Y_{j,t}) + \sum_{j \neq k} b^i_{j,k} Cov_{t-1}(Y_{j,t}, Y_{k,t}) + \epsilon_{i,t}.$$

The elements of the matrix B_i can be interpreted as risk premia: $b^i_{j,j}$ measures the risk premium on the expected return of asset i following a shock on the volatility of asset j. The coefficients $b^i_{j,k}$ are co-risk premia and measure the effect of changes in covolatilities.

In such a model, $F_t = \Sigma_t$ is the unobserved factor. In this case, the factor value is a symmetric positive definite matrix. We could also define the factor as the vector obtained by stacking the different elements of symmetric matrix Σ_t in a vector, that is $F_t = vech(\Sigma_t)$. In practice, the matrix representation of the factor is generally chosen.

Specification [5.7] is directly written as a regression model, with nonlinear effect of $F_t = \Sigma_t$. This specification is

---

is also equal to $\sum_{i=1}^n \sum_{j=1}^n b_{i,j} a_{i,j}$, if matrix A is symmetric. Under the symmetry condition satisfied for instance by $A = \Sigma_t$, it provides the sum of the product of elements of the two matrices.

equivalently written in terms of the transition density as:

$$l(y_t|\, y_{t-1}, \Sigma_t; C, B_1, ..., B_n) =$$

$$\frac{1}{(2\pi)^{n/2}(\det \Sigma_t)^{1/2}} \exp\left\{-\frac{1}{2}\left[Y_t - CY_{t-1} - \begin{bmatrix} Tr(B_1\Sigma_t) \\ \vdots \\ Tr(B_n\Sigma_t) \end{bmatrix}\right]' \Sigma_t^{-1}\right.$$

$$\left.\left[Y_t - CY_{t-1} - \begin{bmatrix} Tr(B_1\Sigma_t) \\ \vdots \\ Tr(B_n\Sigma_t) \end{bmatrix}\right]\right\}. \qquad [5.8]$$

In terms of the conditional Laplace transform, we get:

$$E\left[\exp(z'Y_t)\middle|\, Y_{t-1}, \Sigma_t; C, B_1, ..., B_n\right] =$$

$$\exp\left\{z'\left(CY_{t-1} - \begin{bmatrix} Tr(B_1\Sigma_t) \\ \vdots \\ Tr(B_n\Sigma_t) \end{bmatrix}\right) + \frac{z'\Sigma_t z}{2}\right\}. \qquad [5.9]$$

As in the Gaussian linear dynamic model, the logarithm of the conditional Laplace transform is an affine function of the conditioning variables, that are Y_{t-1} and Σ_t.

5.2.2. *Specification of the volatility dynamics*

The standard family of distributions for stochastic symmetric positive definite matrices is the Wishart family of distributions [PRE 82]. It can be applied to either the conditional distribution of Σ_t given Σ_{t-1}, leading to Wishart autoregressive specifications, or to the conditional distribution of Σ_t^{-1} given Σ_{t-1}, leading to inverse Wishart autoregressive specifications. Wishart autoregressive models are appropriate for Laplace transform-based analysis with estimation by appropriate method of moments. Inverse Wishart autoregressive models are appropriate for density-based analysis with Bayesian estimation of parameters.

In the one dimentional case, $n = 1$, the Wishart distribution coincides with the Gamma distribution. The Wishart autoregressive specification simplifies to an autoregressive Gamma (ARG) process [GOU 06], which is the time discretized Cox, Ingersoll, Ross process [COX 85].

5.2.2.1. *Wishart and inverse Wishart distributions*

The Wishart and inverse Wishart distributions are continuous distributions for stochastic symmetric positive definite matrix, say Σ. The Wishart and inverse Wishart families depend on three parameters, that are degree of freedom $\nu > n - 1$, where (n, n) is the dimension of Σ, a scale matrix S, which is a (n, n) deterministic symmetric positive definite matrix, and a noncentrality parameter, which is also a (n, n) matrix. These two families are related in the following way, in the simplest case of central Wishart distribution.

DEFINITION 5.1.– Σ *follows the inverse (central) Wishart distribution:* $\Sigma \sim IW(\nu, S)$, *if and only if* $\Omega \equiv \Sigma^{-1}$ *follows the central Wishart distribution* $\Omega \equiv \Sigma^{-1} \sim W(\nu, S^{-1})$.

The choice between the Wishart and inverse Wishart families depends on the selected modeling strategies. Indeed the inverse central Wishart distribution has a closed form density, but a complicated Laplace transform (see e.g. [PHI 06, FOX 11, GOL 12]), whereas the central as well as the noncentral Wishart distributions have simple Laplace transforms, but complicated densities (see e.g. [GOU 09]).

1) Inverse central Wishart distribution

The p.d.f. of the Inverse central Wishart distribution $IW(\nu, S)$ is:

$$g(\Sigma; \nu, S) = \frac{|\det S|^{\nu/2}}{2^{(\nu n)/2}\Gamma_n(\nu/2)} |\det \Sigma|^{-\frac{\nu+n+1}{2}} \exp\left[-\frac{1}{2}Tr(S\Sigma^{-1})\right], \quad [5.10]$$

where Γ_n is the multivariate gamma function, introduced to ensure that the measure has unit mass[3].

In the one dimentional case, the central Wishart and inverse central Wishart distributions coincide with Gamma and inverse Gamma distributions, respectively. The p.d.f. of the inverse Gamma $I\gamma(\nu, s)$ is:

$$g(f; \nu, s) = \frac{s^{\nu/2}}{2^{\nu/2}\Gamma(\nu/2)} f^{-\frac{\nu}{2}-1} exp\left(-\frac{1}{2}\frac{s}{f}\right), f \geq 0, \quad [5.11]$$

where $\Gamma(\nu) = \int_0^\infty exp(-f)f^{\nu-1}df$.

2) Noncentral Wishart distribution

The Laplace transform of the noncentral Wishart distribution can be written as:

$$E\left\{\exp\left[Tr(V\Sigma)\right]\right\} = \frac{\exp\left\{Tr\left[V(Id-2SV)^{-1}M\right]\right\}}{\left[\det(Id-2SV)\right]^{\nu/2}}, \quad [5.12]$$

where ν is the degree of freedom, S the scale matrix and M the noncentrality parameter. The definition of the Laplace transform by means of the expectation of $Tr(V\Sigma)$, where V

3 Thus the gamma function is defined by: $\Gamma(\nu) = \int_{\Sigma \gg 0} \exp(-Tr(\Sigma))(\det \Sigma)^{\nu-\frac{n+1}{2}} d\Sigma$, where the integration is on the set of positive symmetric matrices.

is a matrix argument, is convenient for stochastic matrices. It corresponds to the standard definition, since:

$$Tr(V\Sigma) = \sum_{i=1}^{n}(V\Sigma)_{ii} = \sum_{i=1}^{n}\sum_{j=1}^{n} v_{ij}\sigma_{ij},$$

is another way for considering a linear combination of elements of matrix Σ.

5.2.2.2. *Autoregressive processes based on the Wishart and inverse Wishart distributions*

Alternative autoregressive specifications have been used in the literature for the stochastic volatility dynamics. They are defined from either the Wishart, or the inverse Wishart distribution by making the scale or the noncentrality matrices path dependent, that is, function of $\Sigma_{t-1}, ..., \Sigma_{t-p}$ for an autoregressive order p.

1) The inverse Wishart autoregressive process (IWAR process) (see e.g. [PHI 08, GOL 12])

This specification is based on the inverse central Wishart distribution and assumes that the scale matrix S_t is a linear combination of $\Sigma_{t-1}, ..., \Sigma_{t-p}$:

$$\Sigma_t | \underline{\Sigma_{t-1}} \sim IW\left(\nu, \sum_{j=1}^{p} A_j \Sigma_{t-j} A_j'\right). \qquad [5.13]$$

The matrix coefficients $A_j, j = 1, ..., p$ play the role of autoregressive coefficients. They are introduced in a quadratic way to ensure that the combination $\sum_{j=1}^{p} A_j \Sigma_{t-j} A_j'$ is still a symmetric positive definite matrix. This specification can be rewritten in terms of the matrix $\Omega_t = \Sigma_t^{-1}$ as:

$$\Omega_t | \underline{\Omega_{t-1}} \sim W\left(\nu, \left[\sum_{j=1}^{p} A_j \Omega_{t-j}^{-1} A_j'\right]^{-1}\right). \qquad [5.14]$$

Clearly an alternative definition specification could have been obtained by writing a similar model on the central Wishart distribution directly as:

$$\Sigma_t|\underline{\Sigma_{t-1}} \sim W\left(\nu, \sum_{j=1}^{p} A_j \Sigma_{t-j} A_j'\right), \text{ say.} \qquad [5.15]$$

that is, by using an arithmetic average of lagged volatilities instead of the harmonic average appearing in formula [5.14].

2) The Wishart autoregressive process (WAR process) (see e.g. [GOU 09])

This specification is based on the noncentral Wishart distribution and the lagged volatilities introduced by means of the noncentrality parameter. The associated Laplace transform is:

$$E\left\{\exp\left[Tr(V\Sigma_t)\right]|\underline{\Sigma_{t-1}}\right\} =$$

$$\frac{\exp\left\{Tr\left[V(Id-2SV)^{-1}\sum_{j=1}^{p} A_j\Sigma_{t-j}A_j'\right]\right\}}{\left[\det(Id-2SV)\right]^{\nu/2}}. \qquad [5.16]$$

5.2.3. *Overparametrization and constrained models*

The unconstrained multivariate stochastic volatility models for returns encounter the curse of dimensionality. First, the latent factors are the elements of the volatility–covolatility matrix, which implies a number of latent factors $K = n(n+1)/2$ much larger than the number n of the asset returns. The number of parameters is also very large. For instance, we have $n^2(n + 1)/2$ parameters to represent the different risk premia in the dynamic of returns given the volatility, and many more to characterize the volatility dynamics.

The curse of dimensionality can be solved by introducing restrictions on matrices $C, B_1, \ldots, B_n, S, A_1, \ldots, A_p$. Two types of restrictions are encountered in the applied literature:

1) Some models are constrained by zero restrictions on such matrices. For instance, the equation for the return of asset i is often specified as:

$$Y_{i,t} = c_{i,i} Y_{i,t-1} + b^i_{i,i} V_{t-1}(Y_{i,t}) + \epsilon_{i,t}, i = 1, \ldots, n. \qquad [5.17]$$

Thus, only the i specific log-return and i specific volatility are introduced as explanatory variables. Such models are usually introduced since they are easily estimated by one dimentional methods. However, by breaking the links between assets in the autoregressive equation (except the ones coming from the correlations between the errors, that is from the non diagonal form of Σ_t), we likely misspecify the contagion phenomena.

2) Another possibility is to assume that the lagged returns and the volatility effects are passing by means of a small number of portfolios. This approach extends the CAPM (see [SHA 64, LIN 65]) in which the effects pass by means of the market portfolio. It allows to exhibit the number of portfolios summarizing the information, to distinguish the ones summarizing the information for lagged returns from the ones summarizing the information for the volatilities, and to get interpretations of such benchmark portfolios, which can differ from the standard interpretations in terms of market portfolio. An example of such a constrained model with one underlying portfolio for lagged returns and one underlying portfolio for volatilities could be:

$$Y_{i,t} = \gamma_i c' Y_{t-1} + \delta_i \beta' \Sigma_t \beta + \epsilon_{i,t}. \qquad [5.18]$$

The effects of lagged returns pass through the lagged return of the portfolio with allocation c. They may differ among assets due to the different sensitivity coefficient γ_i. Similarly, the volatility effects are transmitted through the volatility of the portfolio with allocation β. These effects are asset dependent due to the different sensitivity coefficients δ_i.

In the constrained model [5.18], the dimensionality is largely diminished with a number of underlying factors equal to 1 instead of $n(n+1)/2$, and a number of parameters equal to $2n-1$ (n for the δ's, n for the β's, -1 to solve the identification issue) instead of $n^2(n+1)/2$.

5.3. Application to portfolio management

5.3.1. *Information sets*

When considering portfolio management, it is important to distinguish the information of the investor and the information of the econometrician. It is generally assumed that the investor is more informed than the econometrician, that is, the investor knows the current and lagged values of both[4] Y_t, Σ_t, whereas the econometrician knows the current and lagged values of Y_t only. However, the econometrician can reconstitute filtered values $\hat{\Sigma}_t$ of Σ_t based on $\underline{Y_{t-1}}$ by appropriate techniques more complex than the Kalman filter described in the appendix to Chapter 3, section 3.7.3.

Loosely speaking, a mean-variance efficient portfolio has allocation in risky assets proportional to:

$$a_t^* \propto \Sigma_t^{-1} \left[CY_{t-1} - \begin{pmatrix} Tr(B_1\Sigma_t) \\ \vdots \\ Tr(B_n\Sigma_t) \end{pmatrix} \right], \text{ for the investor,}$$

4 For instance, since the investor derives the elements of Σ_t from the prices of derivatives written on the basic assets.

$$a_t^{**} \propto \hat{\Sigma}_t^{-1} \left[CY_{t-1} - \begin{pmatrix} Tr(B_1\hat{\Sigma}_t) \\ \vdots \\ Tr(B_n\hat{\Sigma}_t) \end{pmatrix} \right], \text{ for the econometrician,}$$

where $\hat{\Sigma}_t$ is a function of $\underline{Y_{t-1}}$.

Anyway, such efficient portfolio allocations depend on a number of time varying variables much larger than the number n of assets in unconstrained models. Then a $K-$fund separation theorem cannot be applied (see Cass and Stiglitz [CAS 70] for the condition for fund separation theorem). The same conclusion arises in a constrained model such as [5.18].

5.3.2. *Arbitrage opportunities*

In model with constant volatility matrix Σ, it is usual to consider the singular value decomposition (SVD) of Σ (or its estimate $\hat{\Sigma}$), that is, the set of its eigenvalues and eigenvectors. Then to focus on the eigenvectors associated with very small eigenvalues. Indeed those eigenvectors correspond to portfolio allocations, which are almost riskless. Thus, we can expect (quasi) arbitrage opportunities between these portfolios and the risk-free asset.

The same idea can be followed in models with stochastic volatility. But the smallest eigenvalue of Σ_t (or $\hat{\Sigma}_t$) and the associated eigenvector are now path dependent. There exist random dates at which this eigenvalue is close to zero. The model with such stochastic volatility can be used to predict these dates at which arbitrage opportunities might exist, and to estimate the portfolio allocation allowing such a (quasi) arbitrage.

5.3.3. *Volatility index*

The model can also be used to construct volatility indices in a way similar to the construction of market indices. Let us consider the constrained model [5.18] for illustration, with Rank c = Rank β = 1. The portfolio with allocation c summarizes all effects of past returns (conditional on the volatility) and its returns can be the basis of a stock index. Similarly the portfolio with allocation β has a volatility $\beta'\Sigma\beta$, which summarizes all the volatility effects. It differs from the volatility of the stock index, if β is not proportional to c. Thus, its volatility can be used as a *historical* or *realized* volatility index, since it is constructed from the observation of returns only. It differs from *implied* volatility indexes based on the prices of derivatives, such as the VIX[5], which are computed by averaging standardized prices (called implied volatilities) of European call options written on the SP500 market index.

5.4. Chapter 5 highlights

Models with both frailty and contagion can be extended to a nonlinear framework. They can be specified by means of either the transition density, or nonlinear regressions, or Laplace transforms. The approach is illustrated by the example of models with stochastic volatility. These stochastic volatility models are used for detecting endogenous arbitrage opportunities, or for constructing volatility indexes.

5.5. Appendices

5.5.1. *Proof of Proposition 5.1*

The proof is based on the following lemma:

LEMMA 5.1.– Let us consider a continuous real random variable X with strictly increasing cumulative distribution function $G(x) = P[X \leq x]$ and denote Φ the c.d.f. of the

5 The VIX index is the support of options and futures traded on the Chicago Board Options Exchange (CBOE).

standard normal distribution. Then the variable $U = \Phi^{-1}[F(X)]$ is a standard Gaussian variable.

PROOF.– Indeed we have:

$$\begin{aligned} P[U \leq u] &= P[\Phi^{-1}[F(X)] \leq u] \\ &= P[X \leq F^{-1}[\Phi(u)]] \\ &= F(F^{-1}[\Phi(u)]) \\ &= \Phi(u). \end{aligned}$$

The result follows. □

Equivalently, we can write:

$$X = a(U), \text{ say,}$$

with $a = F^{-1}[\Phi]$.

Let us now consider proposition 5.1 and assume for expository purpose $K = n = 1$.

i) The lemma above can be applied to F_t and to its conditional distribution given (F_{t-1}, Y_{t-1}), i.e. given F_{t-1} (see assumption A.1). This provides the first nonlinear regression $F_t = b(F_{t-1}, v_t)$, where v_t is standard normal. Moreover, since the distribution of v_t does not depend on F_{t-1}, Y_{t-1}, we deduce that v_t is independent on the past values.

ii) The other regression is obtained by applying the lemma to the variable Y_t and to its conditional distribution given F_t, F_{t-1}, Y_{t-1}, or equivalently given F_t, CY_{t-1} (see assumption A.2).

6

An Application of Nonlinear Dynamic Models: The Hedge Fund Survival

This chapter illustrates the measurement challenge of systematic risk and contagion in finance by an application to hedge funds' (HF) survival, with interpretations in terms of funding and market liquidity risks (see [DAR 14b]). If HFs allow to pool capital from different investors and invest on different markets, they differ from mutual funds and pension funds in different ways. First, their use of leverage or derivatives is not capped by regulators. Their name refers to the hedging techniques used to reduce directional market exposures (often called "beta") and generate returns uncorrelated to market returns (often called "alpha"). Thus, many HF investment strategies aim to achieve positive returns whatever the market conditions are. These strategies are often called absolute return strategies. Second, HF can invest in relatively illiquid assets even if their open-end structure allows liquid additions, or withdrawals by their investors, generally on a monthly or quarterly basis. This can generate a liquidity mismatch between the asset and the liability sides of the fund, and create a liquidity risk for investors. In that sense, HF play the role of banks in the liquidity transformation in the "shadow banking" system. Third, we observe in practice a significant number of HF liquidations that correspond to the decision by the HF manager to stop the management of the fund.

The understanding of the systematic patterns of dependence between the individual liquidation risks in the hedge fund industry is crucial for financial market regulation (see [FIN 13]). The supervisors have to monitor both the funding and market liquidity risks. They may modify and control the funding liquidity exposure by means of restrictions on the use of leverage, the redemption frequency and the minimal requirements for investing in a HF. Supervisors may also limit the market liquidity exposure.

The stress on the current value of the frailty (shocks on systematic funding liquidity risk) and the stress on the contagion matrix (shocks on the speed of propagation) can be accommodated. The analysis of the common risk factors and contagion effects is the first step in the assessment of the impact of the HF industry on systemic risk for the global financial markets and the possibility of cascades into a global financial crisis.

The liquidation of an HF is a qualitative event, which explains why observations of such liquidations cannot be well represented by means of linear modeling, assuming implicitly Gaussian distributions. More precisely, for individual liquidation observations, the distributions of these 0-1 variables (conditional on the common factor and past observations) are necessarily Bernoulli distributions. In the application below, the qualitative data are preliminary aggregated by management style, leading to a Poisson model for aggregate counts.

First we describe the available data, then we introduce a dynamic Poisson model with common frailty and contagion. This model is estimated on HF liquidation data and the results, that are the estimated network, the underlying factor and the sensitivities w.r.t. factor, are interpreted in terms of liquidity. Finally, we examine the consequences of stresses applied to the model.

6.1. HF liquidation data

The HF liquidation data are obtained from the Lipper TASS database[1]. They concern a total of 3799 funds on the period from October 1992 to June 2009. HF styles are generally classified among three major categories: directional, semi-directional and relative value (arbitrage). However, HF strategies within these categories entail different risk and return profiles. Thus, this first classification is generally extended to obtain a more detailed description of the HF strategies. In our empirical analysis, we use the following 9 styles provided by Lipper TASS: long/short equity hedge (LSE), event driven (ED), managed futures (MFs), equity market neutral (EMN), fixed income arbitrage (FI), global macro (GM), emerging markets (EM), multi-strategy (MS) and convertible arbitrage (CONV). The elements contributing to each HF strategy include the HF's investment approach to the market (discretionary/ quantitative), the instruments used (stocks/derivatives), the method used to select investments (top-down/bottom-up) and the level of diversification within the fund. The HF's prospectus offers investors information about these aspects, but the main source of information remains the HF managers' self-declaration made when they voluntary disclose HF returns in commercial databases. Misclassifications occur since it is difficult to control this information. Moreover, some styles such as MS can gather very different investment strategies. Nevertheless, we observe between strategies huge differences in liquidity risk exposures, both on the market side, i.e. the liquidity level of

1 Tremont Advisory Shareholders Services. Further information about this database is provided at www.lipperweb.com/products/LipperTASS.aspx.

the portfolio invested in market instruments and the funding side, i.e. the liquidity conditions offered to investors. The distribution by management style of alive and liquidated funds in the database is reported in Table 6.1.

		Alive funds	(%)	Liq/ funds	(%)	Total	(%)
CONV	Convertible arbitrage	45	2.0%	66	4.30%	*111*	*2.9%*
EM	Emerging markets	227	10.0%	111	7.30%	*338*	*8.9%*
EMN	Equity market neutral	126	5.5%	139	9.10%	*265*	*7.0%*
ED	Event driven	216	9.5%	129	8.50%	*345*	*9.1%*
FI	Fixed income arbitrage	95	4.2%	75	4.90%	*170*	*4.5%*
GM	Global macro	162	7.1%	102	6.70%	*264*	*6.9%*
LSE	Long/short equity	885	38.8%	546	35.90%	*1,431*	*37.7%*
MF	Managed futures	224	9.8%	230	15.10%	*454*	*12.0%*
MS	Multi-strategy	299	13.1%	122	8.00 %	*421*	*11.1%*
	Total	*2,279*	*100.0%*	*1,520*	*100.00%*	*3,799*	*100.0%*

Table 6.1. *Distributions of alive funds in June 2009, and funds liquidated prior to June 2009, for the nine management styles (from [DAR 14b])*

The largest management style in the database of alive and liquidated funds is LSE hedge (about 40%), followed by MFs, MS and ED (each about 10%).

Figure 6.1 displays the variation of the liquidation rates over time for the different management styles. This figure shows liquidation clustering both with respect to time and among management styles. One liquidation clustering is due to the long-term capital management (LTCM) debacle, observed in the summer of 1998 and linked with the Russian crisis; it is especially visible for the EM and GM styles. Another liquidation clustering is observed in the 2008 financial crisis, but did not concern the GM style.

6.2. Dynamic Poisson model

In each month t and for each management style $k = 1, ..., K$, we observe the number $n_{k,t}$ of HF alive at the beginning of the month, and the liquidation count $Y_{k,t}$ during the month. The value of the unobservable common factor for month t is denoted by F_t. We specify the model in two steps. First, we specify the conditional distribution of the vector $Y_t = (Y_{1,t}, ..., Y_{K,t})'$ of liquidation counts given the current and past factor values $\underline{F_t} = (F_t, F_{t-1}, ...)$ and the past liquidation counts $\underline{Y_{t-1}} = (Y_{t-1}, Y_{t-2}, ...)$. Second, we specify the conditional distribution of the factor value F_t given the past histories $\underline{F_{t-1}}$ and $\underline{Y_{t-1}}$. For expository purposes, we consider a single factor.

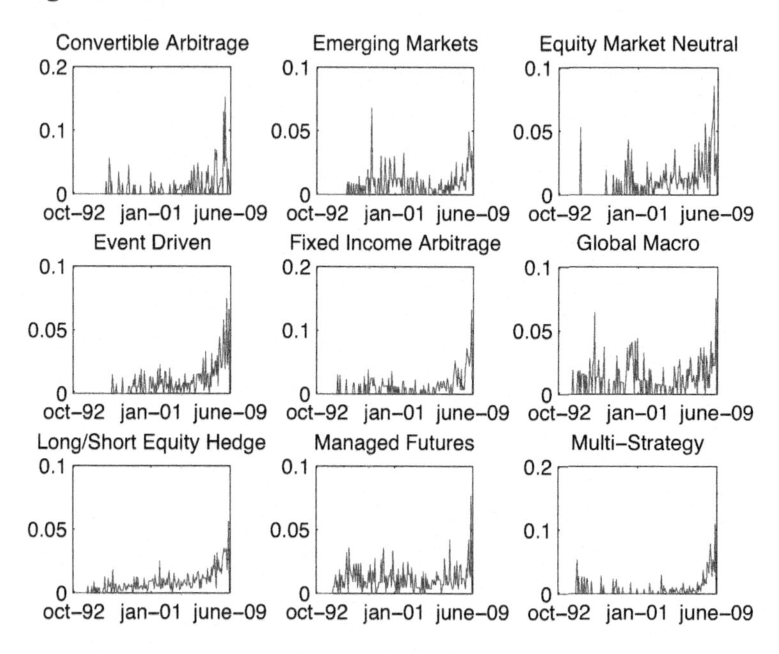

Figure 6.1. *Liquidation rates of HF. The figure displays the time series of monthly liquidation rates between October 1992 and June 2009 for the nine management styles (from [DAR 14b])*

6.2.1. *Conditional distribution of the liquidation counts*

MEASUREMENT EQUATION.– The liquidation counts $Y_{k,t}$ are conditionally independent across management styles given $\underline{F_t}$

and $\underline{Y_{t-1}}$, with Poisson distributions:

$$Y_{k,t}|\underline{F_t}, \underline{Y_{t-1}} \sim \mathcal{P}\left[\gamma_{k,t}(a_k + b_k F_t + c_k' Y_{t-1}^*)\right],$$
$$\text{independent across } k = 1, ..., K \qquad [6.1]$$

where $Y_t^* = (Y_{1,t}/n_{1,t}, ..., Y_{K,t}/n_{K,t})'$ is the vector of liquidation rates, a_k and b_k are the scalar coefficients, c_k is a vector of coefficients of dimension K, and $\gamma_{k,t} = n_{k,t}/n_{k,t_0}$.

Thus, the liquidation intensity includes two components. The first component, $a_k + b_k F_t$, is the intensity of being liquidated via the exogenous factor represented by F_t and the vector $b = (b_1, ..., b_K)'$ of exposures to this factor. The second component, $c_k' Y_{t-1}^* = \sum_{l=1}^{K} c_{k,l} Y_{l,t-1}^*$, is the intensity to be liquidated via the contact with previously liquidated funds, in the same or in a different management style. The contagion is introduced with a lagged effect to capture the propagation phenomenon.

The model is adjusted (1) for the time-varying sizes of the subpopulations, via the term $\gamma_{k,t} = n_{k,t}/n_{k,t_0}$ scaling the baseline intensity and (2) for the density of risky HF in the management style, via the use of rates Y_{t-1}^* instead of counts Y_{t-1} as explanatory variable. To ensure the positivity of the liquidation intensity, we assume that the frailty process F_t is positive valued, and that the coefficients a_k, b_k and $c_{k,l}$ are non-negative.

The contagion effect is measured by means of the $K \times K$ contagion matrix C with rows c_k', $k = 1, ..., K$. This modeling enables us to account for contagion within as well as between management styles, since both diagonal and non-diagonal elements of the contagion matrix C can be non-zero.

6.2.2. *The common factor dynamics*

We complete the model by specifying the conditional distribution of the common factor process. We assume that the frailty variable follows an autoregressive gamma (ARG) process. The ARG process is the time-discretized Cox, Ingersoll, Ross process [COX 85]. The transition of this Markov process corresponds to a non-central gamma distribution $\gamma(\delta, \eta F_{t-1}, \nu)$, where $\nu > 0$ is a scale parameter, $\delta > 0$ is the degree of freedom of the gamma transition distribution and parameter $\eta \geq 0$ is such that $\rho = \eta\nu$ is the first-order autocorrelation (see section 6.6.1 for basic results on the ARG process). Since the factor is unobservable, it is always possible to assume $E(F_t) = 1$ for identification purposes. Thus, the frailty dynamics can be conveniently parameterized by parameters δ and ρ as stated in the following.

STATE EQUATION.– The conditional distribution of F_t given $\underline{F_{t-1}}$ and $\underline{Y_{t-1}}$ depends on F_{t-1} only and is non-central gamma:

$$F_t | \underline{F_{t-1}}, \underline{Y_{t-1}} \sim \gamma\left(\delta, \frac{\rho\delta}{1-\rho} F_{t-1}, \frac{1-\rho}{\delta}\right), \qquad [6.2]$$

where $\delta > 0$ and $0 \leq \rho < 1$.

The ARG specification has several advantages. First, it ensures positive values for the factor, and thus a positive intensity, if $b_k \geq 0$ and $c_{k,l} \geq 0$ for all k, l. Second, the joint model [6.1]–[6.2] for the liquidation counts and the factor is an affine model. This facilitates the analysis of the term structure of liquidation rates.

6.3. Results

The pure contagion model, which is the model without frailty effect, is simply:

$$Y_{k,t} \sim \mathcal{P}\left[\gamma_{k,t}(a_k + c_k' Y_{t-1}^*)\right], \quad k = 1, ..., 9. \qquad [6.3]$$

Since the explanatory variable Y_{t-1}^* is observable, this Poisson autoregressive model is easily estimated by maximum likelihood [CAM 98]. For a model with both unobservable common frailty and contagion, it is necessary to integrate out the unobservable factor path. This implies a very complicated expression of the likelihood function, which involves a multi-dimensional integral of very large dimensions. However, consistent estimation moment methods can be developed by using the affine property of the process (see [DAR 14b]). We describe below some estimation results and discuss the interpretations in terms of liquidity risk.

6.3.1. *Estimated networks*

For the constrained and unconstrained models, we estimate the contagion matrix and derive the associated networks after eliminating the non-significant contagion channels (see Figures 6.2 and 6.3). Let us first discuss the network with pure contagion model (Figure 6.2). All strategies seem interconnected either directly or indirectly through multi-step contagion channels. A contagion scheme featuring this property would correspond to a complete structure in Allen and Gale's [ALL 00] terminology.

However, the fact that the two estimated networks are very different indicates that the pure contagion model is misspecified, and that a common unobservable variable has actually to be introduced. This also shows the consequence of this misspecification in terms of networks. The effects of the omitted common frailty create the misleading impression of a complete network with a lot of links between different styles. When a common frailty is introduced, the contagion scheme is clarified.

The contagion occurs along specific directions, such as MS $\to$ EMN $\to$ ED $\to$ FI $\to$ EM, without any evidence of contagion in the reverse direction. Let us carefully discuss this propagation scheme. The management styles can be classified into four categories: (1) funds mainly invested in fixed income products and using high leverage, which are FI, MF, EM and GM; (2) funds mainly invested in equities, such as EMN, LSE hedge and ED; (3) funds in the CONV management style, in which the convertible products have features of both the corporate bonds and associated stocks; (4) funds in the MS management style, with portfolios including subprimes and equities.

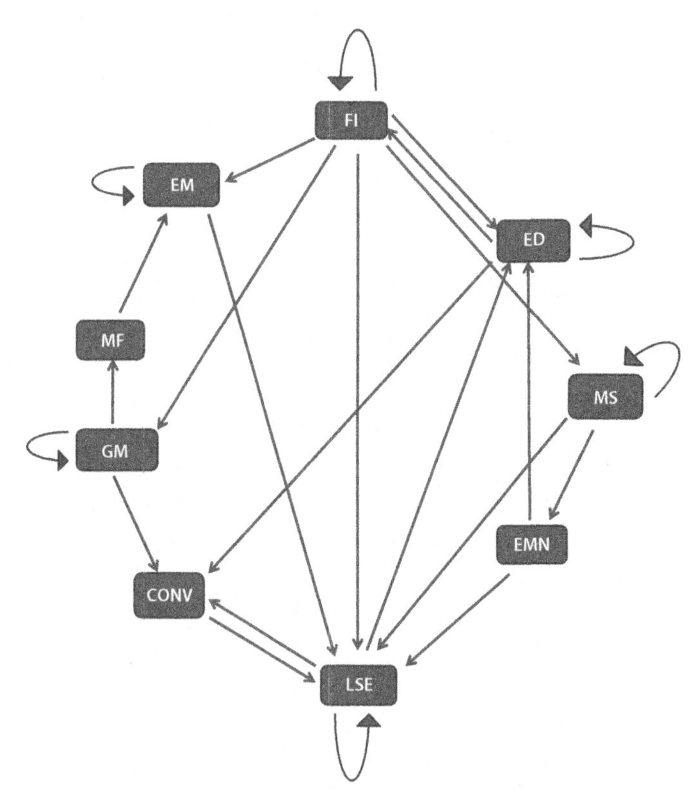

Figure 6.2. *The contagion scheme for the pure contagion model. Arrow between two management styles indicates that the estimated contagion coefficient from the first style to the second style is statistically significant at 5% level (from [DAR 14b])*

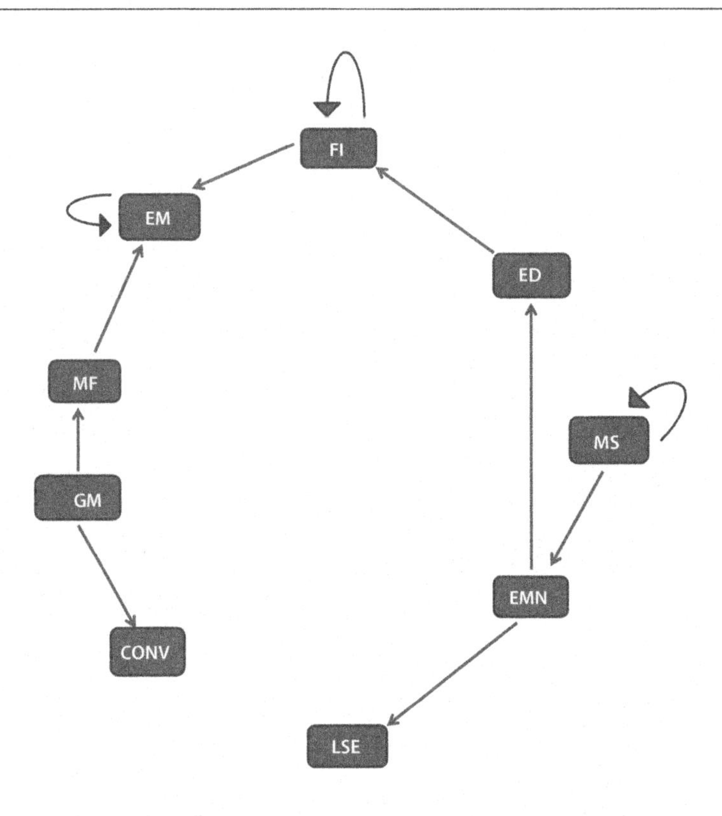

Figure 6.3. *The contagion scheme for the model with contagion and frailty. Arrow between two management styles indicates that the estimated contagion coefficient from the first style to the second style is statistically significant at 5% level (from [DAR 14b])*

Let us now comment on the propagation scheme in relation to the recent subprime crisis and the associated lack of market liquidity in various classes of assets. In 2007, there has been an increase in expected default rates for mortgages, followed by an increase in margin calls for credit derivatives. The MS funds, which were exposed to subprimes, needed cash in order to satisfy the margin requirements. Thus, they had to liquidate the most liquid part of their portfolios, i.e. the equities. This massive deleveraging had a direct effect on the stock prices, increasing the market liquidity risk. At the beginning, this effect was not observed in the stock indices, but mainly in the relative performances of individual stocks: the high-ranked stocks becoming low ranked and vice versa (see [KHA 11]). This dislocation effect of stock prices has impacted all the equity strategies, including the ED funds.

The associated Mergers & Acquisitions (M&A) strategies have transformed the short-term shocks into long-term shocks. This explains the key (systemic) role of the ED management style, which creates the link between the shocks to stock markets and the shocks to fixed income markets. This discussion highlights the key role of the management style ED for the transmission of the liquidity risk from the equity market to the bond market. This is an example of style "too interconnected to fail". Note that the LSE hedge management style is the largest in our dataset, but does not play a central role in the contagion scheme. This confirms the idea that systemic risk is not necessarily associated with size ("too big to fail").

6.3.2. *Factor and factor sensitivities*

6.3.2.1. *Factor*

The underlying factor can be estimated. Its evolution is provided in Figure 6.4. The frailty features a rather stable path between 1996 and 2006, with spikes at the end of 2001 (the 9/11 terrorist attack), the end of 2002 (the market confidence crisis due to the Internet bubble). The frailty path features an upward trend over the years 2007 and 2008 (the recent financial crisis), and decreases rapidly afterward.

This factor is linked to standard measures of funding liquidity risk, which are the Treasury-Eurodollar (TED) spread, the volatility index (VIX) and the spread (SPR) between the US BAA and AAA rated yields. The TED spread is a measure of the refinancing cost on the clearing houses. It is introduced to capture a part of the rollover funding liquidity risk. The VIX is a weighted average of the implied volatility in the Standard & Poor's (S&P) index options. This index measures the aggregate volatility of the stock market as well as the price of this volatility. The spread SPR is obtained from the Federal Reserve (FED) database at the Federal Reserve Bank of St. Louis (see Figure 6.5, lower panel). It is often stated that part of this spread is unrelated to credit risk and is due to the lower liquidity of the corporate bonds in the more risky rating classes.

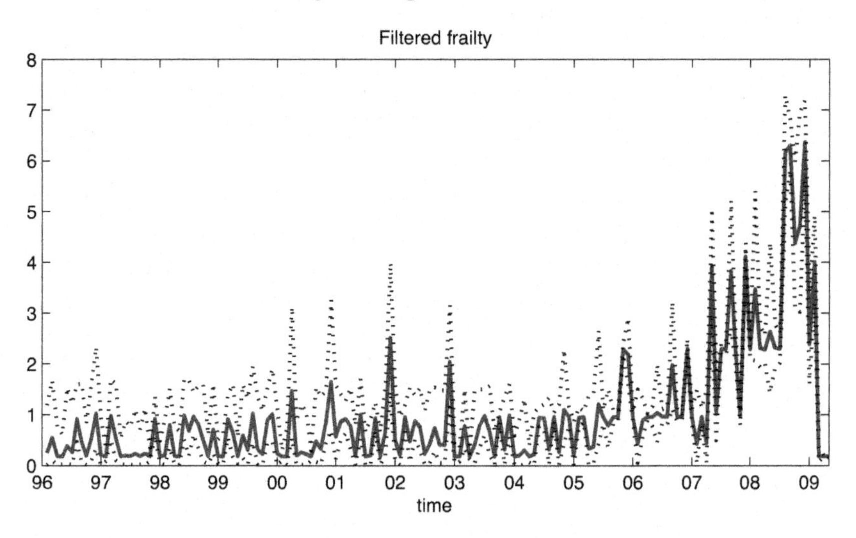

Figure 6.4. *Filtered path of the frailty. The figure displays the filtered path of the frailty (solid line) and the pointwise 95% confidence bands (dotted lines) between January 1996 and June 2009. The filtered value (respectively, the lower and upper confidence bands) at a given month is the median (respectively, the 2.5 and the 97.5% quantiles) of the filtering distribution (from [DAR 14b])*

It has been shown in [DAR 14b] that the link between this factor and these standard measures of funding liquidity is not linear, but features two regimes. This can be understood in terms of equilibria. This occurs in the standard situation of reasonable funding liquidity costs. In the "good equilibrium", the HFs provide liquidity and are invested in rather illiquid assets with high leverage. But, as noted in [BEN 12], when the refinancing costs increase, HFs reallocate their portfolios, reduce their equity holdings and try to diminish their leverage in order to anticipate the consequences of possible outflows. In this "bad equilibrium", the HFs are liquidity seekers.

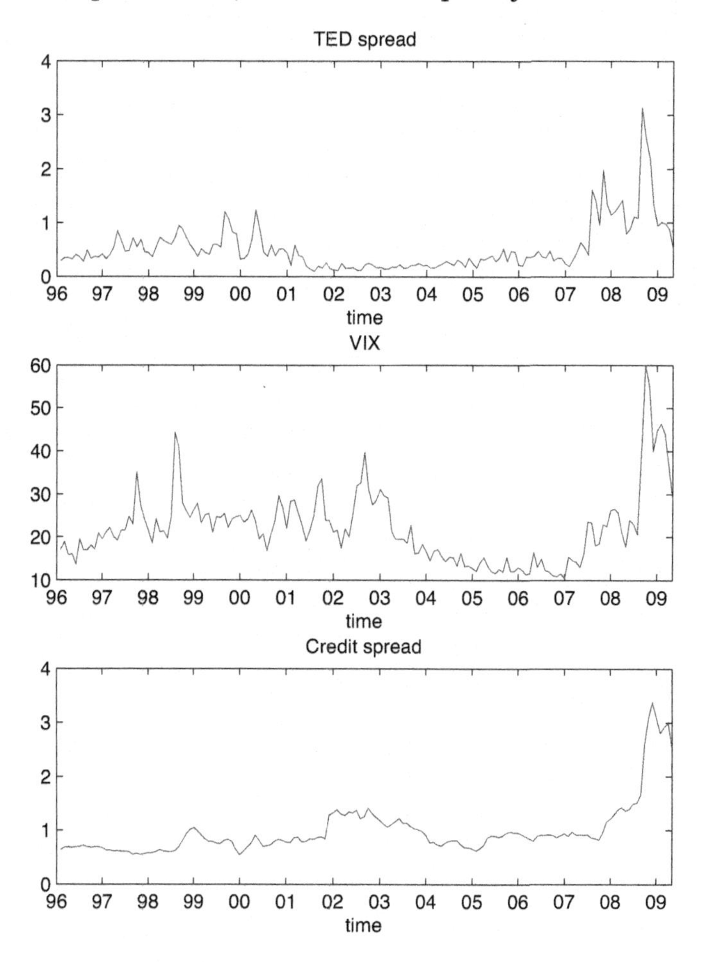

Figure 6.5. *Funding liquidity indicators. The figure displays the monthly time series of the Treasury-Eurodollar (TED) spread, the volatility index (VIX) and the credit spread, measured as the difference between the BAA and AAA yields, between January 1996 and June 2009. The three series are in percentage (from [DAR 14b])*

6.3.2.2. *Factor sensitivities*

The estimated factor sensitivities are given in Table 6.2. The interpretation of the dynamic frailty as a measure of funding liquidity risk is supported by the analysis of these estimated sensitivities. For a given management style, the liquidity features are twofold: the portfolio can be invested in more or less liquid assets (i.e. assets without a large haircut in the case of fire sales), and the strategy can require a longer or shorter horizon to be applied. In this respect, GM and MF portfolios are invested in liquid assets; they can offer to investors weekly, or even daily liquidity conditions, and have small sensitivity coefficients 0.33 and 0.63, respectively. At the opposite, the ED strategies are essentially looking for positive outcomes in mergers and acquisitions, which can only be expected in a medium horizon. They have less interesting, generally quarterly, liquidity conditions, and the factor sensitivity coefficient 1.39 is the second highest one. Similar remarks can be done for other management styles.

The discussion above is completed by comparing the factor sensitivities with the average redemption frequencies and leverage in each management style, which are displayed in Table 6.3.

While the majority of funds across management styles allow for redemptions on a monthly basis or more often, we observe some differences, especially in the ED management style, where the redemption frequency is often close to 3 months. It has been observed that "hedge funds with favorable redemption terms differ significantly in terms of their appetites for liquidity risks" (see [TEO 11]).

In Table 6.3, we observe a significant negative link between the factor sensitivity and the proportion of HFs with a redemption frequency of 1 month or less (respectively, of HF reporting the use of leverage). This link is likely explained by the type of assets introduced in the HF portfolio. For instance, as already remarked, the funds in the management styles MF and GM are invested in very liquid assets. Thus, they can easily propose good redemption frequencies and use leverage without being too sensitive to the common factor. However, to attract investors, some managers in other strategies may propose "favorable" redemption conditions and simultaneously post high returns obtained by taking an excessive liquidity risk. This is likely the case for some HFs in the LSE category, where the favorable announced redemption and the usual high leverage are not in line with the very high exposure to the funding liquidity risk factor.

	Sensitivity b_k
Convertible arbitrage	1.08^{**}
	(0.55)
Emerging markets	0.69^{**}
	(0.27)
Equity market neutral	0.84^{**}
	(0.40)
Event driven	1.39^{**}
	(0.70)
Fixed income arbitrage	0.31^{**}
	(0.13)
Global macro	0.33^{***}
	(0.12)
Long/short equity hedge	4.55^{**}
	(1.92)
Managed futures	0.63^{**}
	(0.25)
Multi-strategy	0.89^{**}
	(0.41)

Table 6.2. *Estimated factor sensitivities in the model with contagion and frailty (from [DAR 14b])*

The link between the frailty sensitivities and the redemption frequencies is confirmed by the correlation between the former and the proportion of favorable redemption conditions (less than 1 month) equal to -0.27, passing to −0.56 when the LSE category is not considered. Similarly, the correlation between the frailty sensitivities and the proportions of funds reporting the use of leverage is −0.32. To summarize, the sensitivity coefficients measure the funding liquidity risk exposures of the different management styles, and these exposures are related to the management of gates and leverage.

	Red. freq. $\leq$ 1 m	1 m $<$ Red. freq. $\leq$ 3 m	Red. freq. $>$ 3 m	Leverage	Sensitivity b_k
Convertible arbitrage	49%	47%	4%	77%	1.08
Emerging markets	66%	31%	3%	57%	0.69
Equity market neutral	68%	28%	4%	58%	0.84
Event driven	36%	48%	15%	54%	1.39
Fixed income arbitrage	46%	48%	6%	68%	0.31
Global macro	80%	18%	2%	70%	0.33
Long/short equity	56%	37%	7%	57%	4.55
Managed futures	92%	7%	1%	79%	0.63
Multi-strategy	67%	29%	4%	50%	0.89

Table 6.3. *Redemption frequency, leverage and factor sensitivity. The second, third and fourth columns display the percentages of hedge funds with redemption frequency smaller or equal to 1 month, between 1 and 3 months, and larger than 3 months, respectively, for the nine management styles. The fifth column displays the percentage of hedge funds reporting some use of leverage. For comparison purposes, the last column of the table provides the sensitivities to the frailty estimated by GMM in the Poisson's model with frailty and contagion (see Table 6.2) (from [DAR 14b])*

6.3.3. *The relative importance of frailty and contagion*

The relative effect of contagion and frailty on liquidation risk can be measured by using the variance decomposition given in the next proposition.

PROPOSITION 6.1 ([DAR 14b]).– The variance-covariance matrix of the liquidation count vector Y_t can be decomposed as:

$$V(Y_t) = diag[E(Y_t)] + CV(Y_t)C' + (1/\delta)bb'$$
$$+(1/\delta)\rho C(Id - \rho C)^{-1}bb' + (1/\delta)\rho bb'(Id - \rho C')^{-1}C',$$

where $diag[E(Y_t)]$ denotes the diagonal matrix with diagonal elements corresponding to the elements of the vector of expected liquidation counts $E[Y_t] = (Id - C)^{-1}(a + b)$, and a and b denote vectors with elements a_k and b_k, respectively.

In the decomposition of the historical variance-covariance matrix of the liquidation counts, the first term $diag[E(Y_t)]$ on the right-hand side is the variance in a Poisson's model with cross-sectional independence. The sum of the first and second terms provides the expression of the variance in a model including contagion, but without frailty. The third term $(1/\delta)bb'$ captures the direct effect of the exogenous frailty. The remaining terms accommodate its indirect effects through contagion, namely, the amplification of the frailty effect due to the network. This variance decomposition is written in an implicit form as the system in proposition 6.1 has to be solved to get the expression of $V(Y_t)$ as a function of the model parameters.

Let us now assess the magnitude of the terms in the variance decomposition by using the estimated model with contagion and frailty. To do that, we consider a portfolio of liquidation swaps (LSs) written on the individual HFs, which is diversified with respect to the management styles. The LS for management style k pays 1 USD for each fund of style k that is liquidated in month t. By using the variance decomposition, we can evaluate the percentage of portfolio variance due to the underlying idiosyncratic (i.e. management style-specific) Poisson shocks, contagion and frailty, respectively. The decomposition is displayed in Table 6.4.

The largest contribution to the portfolio variance comes from the frailty process, either through a direct effect (64.30%) or through an indirect effect via the contagion network (24.06%). The remaining part of portfolio variance is explained by the underlying Poisson shocks (6.54%) and the direct contagion effects (5.10%). Even though the direct effect of contagion is modest, the network plays an important role in amplifying the effect of the exogenous frailty.

	Underlying Poisson	Contagion	Frailty (direct effect)	Frailty (propagated by contagion)
Percentage of variance	6.54 %	5.10 %	64.30 %	24.06 %

Table 6.4. *Decomposition of the variance. The table provides the decomposition of the variance of the payoff of an equally weighted portfolio of liquidation swaps. The liquidation swap for management style k pays 1 USD for each fund of style k that is liquidated in a given month (from [DAR 14b])*

6.4. Stress-tests

The estimated model with dynamic frailty and contagion can be used for portfolio management of a fund of funds, or for the computation of reserves. In this section, we illustrate how to implement the prediction of future liquidation counts and the stress-tests for liquidation risk. We consider a portfolio of HF with fixed style sizes and compare the distributions of the future liquidation counts in the unstressed and stressed situations. The future counts are subject to a double uncertainty, which is the drawing of the idiosyncratic risks in the Poisson conditional distribution, and the stochastic evolution of the exogenous dynamic frailty. The analysis in the unstressed situation corresponds to the prediction of any (nonlinear) function of the liquidation counts and serves as a benchmark. In the conditioning set, the unobservable current value F_t of the frailty is replaced by its filtered value (see section 6.3.2.1). The stress can be designed in the following ways:

1) We can stress the current factor value from the filtered value to the 95% quantile of its historical distribution.

2) Alternatively, we can change parameters values, for instance, by "increasing" the matrix of contagion C from $\hat{C}$ to

$2\hat{C}$. This stress scenario increases the liquidation risks by amplifying the impact of the exogenous shock by contagion.

For all stress scenarios, the filtered value of the frailty and the vector of observed liquidation counts in the conditioning set correspond to the last month of the sample, i.e. June 2009. In Figures 6.6 and 6.7, we display the impact of stress scenarios on the term structures of the conditional expectations of liquidation counts for the nine management styles.

In each figure, the squares represent the term structures of the expected liquidation counts before stress, which are the same in each scenario. As the horizon increases, the term structure converges to the unconditional expectation of the liquidation count, for each management style. The unstressed term structures are upward sloping, since the current month, i.e. June 2009, corresponds to a period with few liquidation events in any management style and a small frailty value compared to the historical average. The circles represent the term structures of the expected liquidation counts after the shock.

The two types of shocks have very different effects on the term structures. The shock to the current factor value in the first stress scenario is a transitory funding liquidity shock, with different impacts in the short-term with respect to the management style (see Figure 6.6). Its effect decays rather quickly and disappears after about 12 months.

In the second stress scenario, the change in the contagion matrix is a permanent shock. In Figure 6.7, there is no important effect in the short-term, but the long-term behaviors of the models with and without the shock in the contagion matrix significantly differ for all styles, except for GM. We conclude that there is no significant contagion effect impacting the GM style. Therefore, the stress in the second scenario is irrelevant for the distribution of liquidation counts in that management style (Figure 6.7).

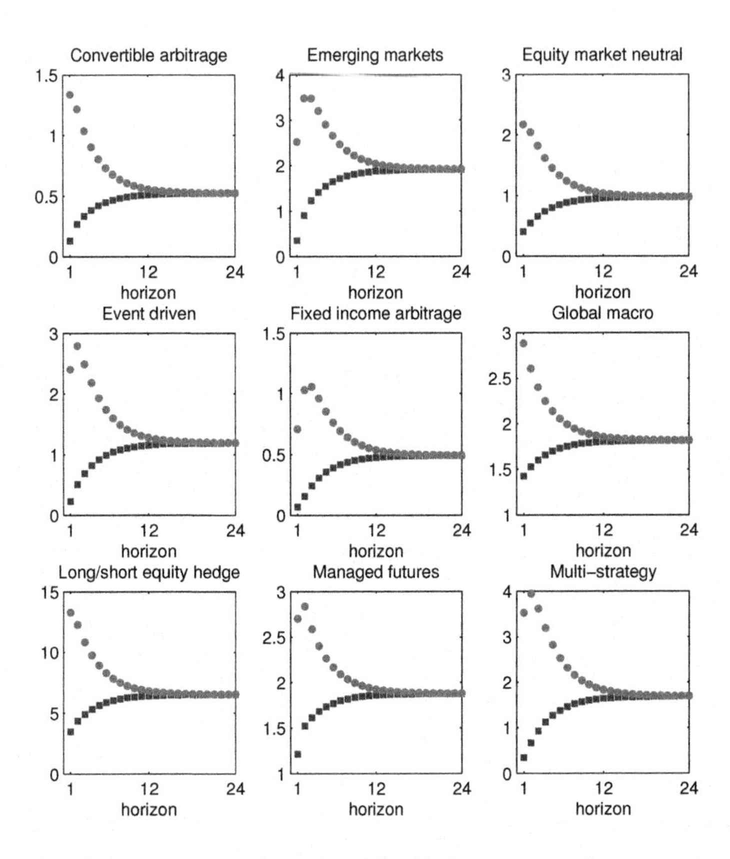

Figure 6.6. *Term structure of expected liquidation counts when stressing the current factor value. Squares and circles correspond to conditioning sets with* F_t *equal to the filtered value of the frailty in June 2009, and the 95% quantile of the stationary distribution of the frailty, respectively. The liquidation counts vector* Y_t *in the conditioning set corresponds to the observations in June 2009 for both curves (from [DAR 14b])*

In the first stress scenario, we obtain an "exogenous crisis" and this can arise without modifying the contagion matrix C, that is the speed of propagation of the shocks. In the second stress scenario, there is a change in the structure and speed of contagion. Such a change can be due to either a change in the behavior of the fund managers or of the supervisor for systemic risk.

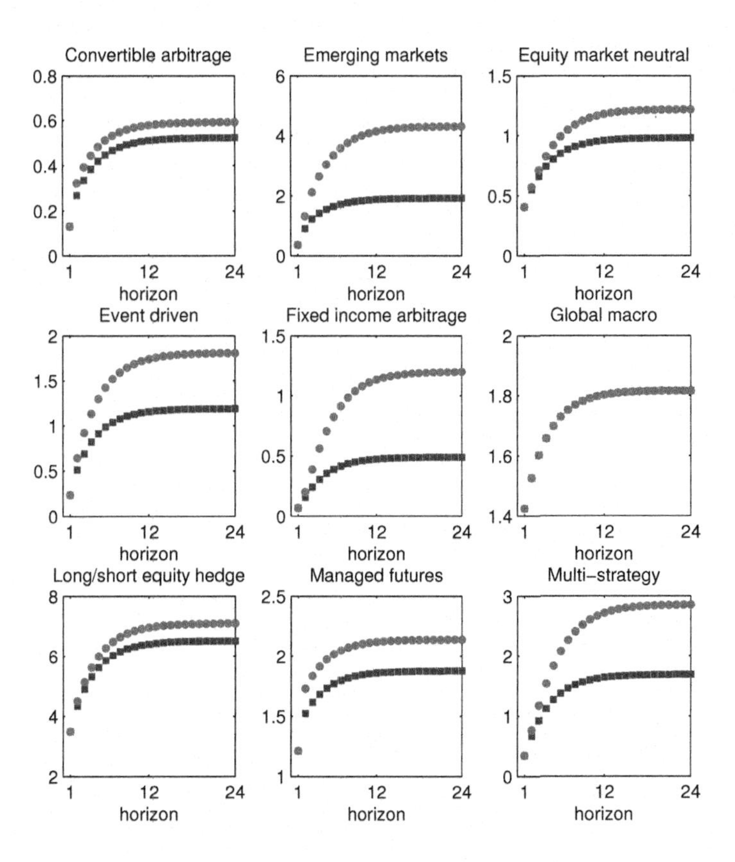

Figure 6.7. *Term structure of expected liquidation counts when stressing the contagion matrix. Squares and circles correspond to models with contagion matrices $\hat{C}$ and $2\hat{C}$, respectively. In the conditioning set, the factor value F_t corresponds to the filtered value of the frailty in June 2009, and the liquidation counts vector Y_t corresponds to the observation in June 2009 (from [DAR 14b])*

To summarize, the model considered in this chapter provides a framework to highlight that the policy maker and the supervisor may have to distinguish between exogenous crises, due to exogenous shocks, and endogenous crises, due to changes in the contagion matrix. To diminish the probability of an exogenous crisis, the policy maker and the supervisor have to control the extreme exogenous risks, that is the distribution of factor F_t.

The usual practice of stress tests is to consider only the exogenous shocks with a given contagion scheme, and to define as carefully as possible the sources of the shocks. For the analysis of systemic risk, it is more relevant to directly shock the underlying common risk factor instead of introducing shocks to the idiosyncratic risks of each management style hidden in the Poisson (conditional) distribution. By selecting the factor, we introduce a common shock to all management styles with different weights given by the frailty beta coefficients.

6.5. Chapter 6 highlights

In this chapter, we develop a methodology to analyze the dynamics of liquidation risk dependence in the HF industry. The autoregressive Poisson model with dynamic frailty is especially convenient as it allows for distinguishing the effects of the exogenous shocks, which affect directly the liability component of the balance sheet, from the endogenous contagion effects, which pass through the asset component. For the application to HF survival, the common factor, the sensitivities to this factor and the contagion scheme can be interpreted in terms of liquidity risks.

The shared dynamic frailty drives the major part of the liquidation clustering in a portfolio of HFs. The direct effect of contagion, that is the transmission of the idiosyncratic shock to an individual HF within and between management styles, is rather limited. However, the contagion scheme has a quantitatively important indirect effect through the amplification of the shocks to the shared frailty.

6.6. Appendices

6.6.1. *The autoregressive gamma process*

In this section, we review the main properties of the ARG(1) process (see [GOU 06a]).

i) *The conditional distribution*

The ARG(1) process (F_t) is a Markov process with conditional distribution the non-central gamma distribution $\gamma(\delta, \eta F_{t-1}, \nu)$, where δ, $\delta > 0$, is the degree of freedom, ηF_{t-1}, $\eta > 0$, is the non-centrality parameter and ν, $\nu > 0$, is a scale parameter. Its first-order and second-order conditional moments are:

$$E(F_t|F_{t-1}) = \delta\nu + \eta\nu F_{t-1}, \quad V(F_t|F_{t-1}) = \nu^2\delta + 2\eta\nu^2 F_{t-1}. \quad [6.4]$$

The ARG(1) process is a discrete-time affine process, that is the conditional Laplace transform is an exponential affine function of the lagged variable:

$$E[\exp(-uF_t)|F_{t-1}] = \exp[-\alpha(u)F_{t-1} - \beta(u)], \quad [6.5]$$

where functions α and β are given by:

$$\alpha(u) = \frac{\eta\nu u}{1+\nu u}, \quad \beta(u) = \delta \log(1 + \nu u), \quad [6.6]$$

for any real value of the argument $u > -1/\nu$.

ii) *The state space representation*

The ARG(1) process admits a state space representation, which is especially convenient for simulating the paths of the process. To get a simulated value of F_t given F_{t-1}, we proceed as follows:

1) We draw an intermediate value Z_t^s in a Poisson distribution $\mathcal{P}(\eta F_{t-1})$.

2) Then, F_t is drawn in the centered gamma distribution $\gamma(\delta + Z_t^s, 0, \nu)$.

iii) *Stationarity condition and stationary distribution*

The ARG(1) process is stationary if $\rho = \nu\eta$ is such that $\rho < 1$. Thus, the stationary distribution is a centered gamma distribution $\gamma(\delta, 0, \frac{\nu}{1-\nu\eta})$. In particular, we get the unconditional moments:

$$E(F_t) = \frac{\nu\delta}{1-\nu\eta}, \quad V(F_t) = \delta\left(\frac{\nu}{1-\nu\eta}\right)^2. \qquad [6.7]$$

From the first equation in [6.4] it is seen that parameter ρ is the first-order autocorrelation of process (F_t).

iv) *Normalization and reparameterization*

When the $ARG(1)$ process is used as a latent frailty, the scale of the process can be absorbed in the sensitivity parameters b_k of the intensity function in [6.1]. Then, the process (F_t) can be normalized to have stationary expectation $E(F_t) = 1$. Thus, from the first equation in [6.7], the parameters are such that $\nu\delta = 1 - \nu\eta = 1 - \rho$. It follows that the model can be parameterized in terms of δ and ρ, while the remaining parameters are given by:

$$\nu = \frac{1-\rho}{\delta}, \qquad \eta = \frac{\rho\delta}{1-\rho}. \qquad [6.8]$$

The stationary distribution is $\gamma(\delta, 0, 1/\delta)$, with Laplace transform $E[\exp(-uF_t)] = (1 + u/\delta)^{-\delta}$, for $u > -\delta$. Moreover, the stationary variance is $V(F_t) = 1/\delta$.

Bibliography

[AKA 11] AKAY O., SENYUZ Z., YOLDAS E., "Understanding hedge fund contagion: a Markov switching dynamic factor model", Working paper, Bentley University, 2011.

[ALE 11] ALESSI L., BARIGOZZI M., CAPASSO M., "Nonfundamentalness in structural econometric models: a review", *International Statistical Review*, vol. 79, pp. 16–47, 2011.

[ALL 00] ALLEN F., GALE D., "Financial contagion", *Journal of Political Economy*, vol. 108, pp. 1–33, 2000.

[AND 84] ANDERSON T., *An Introduction to Multivariate Statistical Analysis*, Wiley, 1984.

[ARS 93] ARSHANAPALLI B., DOUKAS J., "International stock market linkages: evidence from the pre- and post-October 1987 period", *Journal of Banking and Finance*, vol. 17, pp. 193–208, 1993.

[BAE 03] BAE K., KAROLYI A., STULZ R., "A new approach to measuring financial contagion", *Review of Financial Studies*, vol. 16, pp. 717–764, 2003.

[BAL 94] BALL S., ROMA A., "Stochastic volatility option pricing", *Journal of Financial and Quantitative Analysis*, vol. 29, pp. 589–607, 1994.

[BAL 03] BALIOS D., XANTHAKIS M., "International interdependence and dynamic linkages between developed stock markets", *South Eastern Europe Journal of Economics*, vol. 1, pp. 105–130, 2003.

[BAR 13] BARIGOZZI M., BROWNLEES C., "Nets: network estimation for time series", Working paper, University of Pompeu Fabra, 2013.

[BCB 10] BASEL COMMITTEE FOR BANKING SUPERVISION, "Guidance for National Authorities Operating the Countercyclical Capital Buffer", Consultative Document, Bank of International Settlements, Basel, 2010.

[BEN 12] BEN-DAVID I., FRANZONI F., MOUSSAWI R., "Hedge fund stock trading in the financial crisis of 2007-2009", *Review of Financial Studies*, vol. 25, pp. 1–54, 2012.

[BER 05] BERNANKE B., BOIVIN J., ELIASZ P., "Measuring the effects of monetary policy: a factor-augmented vector autoregressive (FAVAR) approach", *Quarterly Journal of Economics*, vol. 120, pp. 387–422, 2005.

[BIL 12] BILLIO M., GETMANSKI M., LO A. *et al.*, "Econometric measures of connectedness and systemic risk in the finance and insurance sectors", *Journal of Financial Economics*, vol. 104, pp. 535–559, 2012.

[BLA 89] BLANCHARD O., QUAH D., "The dynamic effects of aggregate demand and supply disturbances", *American Economic Review*, vol. 78, pp. 655–673, 1989.

[BOY 10] BOYSON N., STAHEL C., STULZ R., "Hedge fund contagion and liquidity shocks", *Journal of Finance*, vol. 65, pp. 1789–1816, 2010.

[BRE 91] BREIDT F., DAVIS R., LII K. *et al.*, "Maximum likelihood estimation for noncausal autoregressive processes", *Journal of Multivariate Analysis*, vol. 36, pp. 175–198, 1991.

[BRU 09] BRUNNERMEIER M., PEDERSEN L.H., "Market liquidity and funding liquidity", *Review of Financial Studies*, vol. 22, pp. 2201–2238, 2009.

[CAM 98] CAMERON A., TRIVEDI P., *Regression Analysis of Count Data*, Cambridge University Press, Cambridge, 1998.

[CAS 70] CASS D., STIGLITZ J., "The structure of investor preferences and asset returns, and separability in portfolio allocation", *Journal of Economic Theory*, vol. 2, pp. 127–160.

[CHA 83] CHAMBERLAIN G., ROTHSCHILD M., "Arbitrage, factor structure and mean-variance analysis in large asset markets", *Econometrica*, vol. 51, pp. 1281–1304, 1983.

[CHA 06] CHAN K., HO L., TONG H., "A note on time reversibility of multivariate linear processes", *Biometrika*, vol. 93, pp. 221–227, 2006.

[CHA 07] CHAN N., GETMANSKY M., HAAS S. *et al.*, "Systemic risk and hedge funds", in CAREY M., STULZ R. (eds), *The Risks of Financial Institutions*, University of Chicago Press, pp. 235–338, 2007.

[COX 85] COX J., INGERSOLL J., ROSS S., "A theory of the term structure of interest rates", *Econometrica*, vol. 53, pp. 385–407, 1985.

[DAR 06] DAROLLES S., GOURIEROUX C., JASIAK J., "Structural Laplace transform and compound autoregressive models", *Journal of Time Series Analysis*, vol. 27, pp. 477–503, 2006.

[DAR 14a] DAROLLES S., DUBECQ S., GOURIEROUX C., "Contagion analysis in the banking sector", Working paper, CREST, 2014.

[DAR 14b] DAROLLES S., GAGLIARDINI P., GOURIEROUX C., "Contagion and systematic risk: an application to the survival of hedge funds", Working paper, CREST, 2014.

[DAV 12] DAVIS R., SONG L., "Noncausal vector AR processes with application to economic times series", Working paper, Columbia University, 2012.

[DIB 88] DIBA B., GROSSMAN H., "Explosive rational bubbles in stock prices?", *American Economic Review*, vol. 78, pp. 520–530, 1988.

[DUF 96] DUFFIE D., KAN R., "A yield factor model of interest rates", *Mathematical Finance*, vol. 6, pp. 379–408, 1996.

[DUF 03] DUFFIE D., FILIPOVIC D., SCHACHERMAYER W., "Affine processes and applications in finance", *Annals of Applied Probability*, vol. 13, pp. 984–1053, 2003.

[DUF 09] DUFFIE D., ECKNER A., HOREL G. *et al.*, "Frailty correlated default?", *Journal of Finance*, vol. 64, pp. 2089–2183, 2009.

[DUN 05] DUNGEY M., FREY R., GONZALEZ-HERMOSILLO B. *et al.*, "Empirical modeling of contagion: a review of methodologies", *Quantitative Finance*, vol. 5, pp. 9–24, 2005.

[FIN 09] FINANCIAL STABILITY BOARD, Guidance to assess the systemic importance of financial institutions, markets and instruments, Internal report, Initial Considerations, 2009.

[FIN 13] FINANCIAL STABILITY BOARD, "Recovery and resolution planning for systematically important institutions: guidance on developing effective resolution strategies", Internal report, July 2013.

[FOR 02] FORBES K., RIGOBON R., "No contagion, only interdependence: measuring stock market co-movements", *Journal of Finance*, vol. 57, pp. 2223–2261, 2002.

[FOX 11] FOX E., WEST M., "Autoregressive models for variance matrices: stationary inverse Wishart processes", Working paper, 2011.

[GOL 12] GOLOSNOY V., GRIBISCH B., LIESENFELD R., "The conditional autoregressive Wishart model for multivariate stock market volatility", *Journal of Econometrics*, vol. 167, pp. 211–223, 2012.

[GOU 82] GOURIEROUX C., LAFFONT J.J., MONFORT A., "Rational expectation in dynamic linear models: analysis of the solutions", *Econometrica*, vol. 50, pp. 409–425, 1982.

[GOU 97] GOURIEROUX C., MONFORT A., *Time Series and Dynamic Models*, Cambridge University Press, 1997.

[GOU 05] GOURIEROUX C., JASIAK J., "Nonlinear innovations and impulse responses with application to VaR sensitivity", *Annales d'Economie et de Statistiques*, vol. 78, pp. 1–31, 2005.

[GOU 06a] GOURIEROUX C., JASIAK J., "Autoregressive gamma process", *Journal of Forecasting*, vol. 25, pp. 129–152, 2006.

[GOU 06b] GOURIEROUX C., MONFORT A., POLIMENIS V., "Affine models for credit risk analysis", *Journal of Financial Econometrics*, vol. 4, pp. 494–530, 2006.

[GOU 07] GOURIEROUX C., MONFORT A., "Econometric specifications of stochastic discount factor models", *Journal of Econometrics*, vol. 136, pp. 509–530, 2007.

[GOU 09] GOURIEROUX C., JASIAK J., SUFANA R., "The Wishart autoregressive process of multivariate stochastic volatility", *Journal of Econometrics*, vol. 150, pp. 107–181, 2009.

[GOU 14] GOURIEROUX C., MONFORT A., PEGORARO F. *et al.*, "Regime switching and bond pricing", *Journal of Financial Econometrics*, vol. 12, pp. 237–277, 2014.

[GOU 15a] GOURIEROUX C., JASIAK J., "Semi-parametric estimation of noncausal vector autoregression", Working paper, CREST, 2015.

[GOU 15b] GOURIEROUX C., MONFORT A., "Revisiting identification and estimation in structural VARMA models", Working paper, CREST, 2015.

[GOU 15c] GOURIEROUX C., ZAKOIAN J.M., "Explosive bubble modelling by noncausal process", Working paper, CREST, 2015.

[GOU 15d] GOURIEROUX C., ZAKOIAN J.M., "On uniqueness of moving average representation of heavy tailed stationary processes", forthcoming in *Journal of Time Series Analysis*, 2015.

[GRA 97] GRANGER C., SWANSON N., "Impulse response functions based on a causal approach to residual orthogonalization in vector autoregression", *Journal of the American Statistical Association*, vol. 92, pp. 357–367, 1997.

[HAN 91] HANSEN L., SARGENT T., "Two difficulties in interpreting vector autoregressions", in HANSEN L., SARGENT T. (eds), *Rational Expectations Econometrics*, Westview Press, Boulder, 1991.

[HES 93] HESTON S., "A closed form solution for option with stochastic volatility with applications to bond and currency options", *Review of Financial Studies*, vol. 6, pp. 327–343, 1993.

[HIG 01] HIGHAM N., "Cholesky factorization", in HAZEWINKEL M. (ed.), *Encyclopedia of Mathematics*, Springer, 2001.

[HUL 93] HULL J., WHITE A., "The pricing of options on assets with stochastic volatility", *Mathematical Finance*, vol. 8, pp. 27–48, 1993.

[JAC 09] JACOBSON N., *Basic Algebra*, 2nd ed., Dover, 2009.

[KAR 96] KAROLYI M., STULZ S., "Why do markets move together? An investigation of U.S.-Japan stock return comovements", *Journal of Finance*, vol. 51, pp. 951–986, 1996.

[KHA 11] KHANDANI A., LO A., "What happened to the quants in August 2007? Evidence from factors and transactions data", *Journal of Financial Markets*, vol. 14, pp. 1–46, 2011.

[KIN 90] KING M., WADHWANI S., "Transmission of volatility between stock markets", *Review of Financial Studies*, vol. 3, pp. 5–322, 1990.

[KOO 96] KOOP G., PESARAN M., POTTER S., "Impulse response analysis in nonlinear multivariate models", *Journal of Econometrics*, vol. 74, pp. 119–147, 1996.

[LAN 13] LANNE M., SAIKKONEN P., "Noncausal vector autoregressions", *Econometric Theory*, vol. 20, pp. 447–481, 2013.

[LAW 71] LAWLEY D., MAXWELL A., *Factor Analysis as a Statistical Method*, Butterworths, 1971.

[LEE 13] LEEPER E., WALKER T., YANG S., "Fiscal foresight and information flows", *Econometrica*, vol. 81, pp. 1115–1145, 2013.

[LIN 65] LINTNER J., "The valuation of risky assets and the selection of risky investments in stock portfolios and capital budgets", *Review of Economics and Statistics*, vol. 48, pp. 13–37, 1965.

[LIP 93] LIPPI M., REICHLIN L., "The dynamic effects of aggregate demand and supply disturbances: comment", *American Economic Review*, vol. 83, pp. 644–652, 1993.

[MAL 92] MALLIARIS A., URRIATA J., "The international crash of october 1987: causality tests", *Journal of Financial and Quantitative Analysis*, vol. 22, pp. 353–364, 1992.

[MAN 93] MANSKI C., "Identification of endogenous social effects: the reflection problem", *Review of Economic Studies*, vol. 60, pp. 531–542, 1993.

[MAR 52] MARKOVITZ H., "Portfolio selection", *Journal of Finance*, vol. 7, pp. 77–91, 1952.

[MAS 98] MASSON P., "Contagion-monsoonal effects, spillovers and jumps between equilibria", Working Paper no. 98/142, IMF, 1998.

[PHI 06] PHILIPOV A., GLICKMAN M., "Multivariate stochastic volatility via Wishart processes", *Journal of Business and Economic Statistics*, vol. 24, pp. 313–328, 2006.

[PRE 82] PRESS S., *Applied Multivariate Analysis*, 2nd edition, Dover Publications, New York, 1982.

[RAM 74] RAMSEY F., "Characterization of the partial autocorrelation function", *Annals of Statistics*, vol. 2, pp. 1296–1301, 1974.

[ROS 76] ROSS S., "The arbitrage theory of capital asset pricing", *Journal of Economic Theory*, vol. 13, pp. 341–360, 1976.

[ROS 00] ROSENBLATT M., *Gaussian and Non-Gaussian Linear Time Series and Random Fields*, Springer-Verlag, 2000.

[SHA 64] SHARPE W., "Capital asset prices: a theory of market equilibrium under conditions of risk", *Journal of Finance*, vol. 19, pp. 425–442, 1964.

[SHA 66] SHARPE W., "Mutual fund performance", *Journal of Business*, vol. 39, pp. 119–138, 1966.

[SIM 77] SIMS C., "Exogeneity and causal ordering in macroeconomic models", in SIMS C. (ed.), *New Methods in Business Cycle Research*, Federal Reserve Bank of Minneapolis, 1977.

[SIM 80] SIMS C., "Macroeconomics and reality", *Econometrica*, vol. 48, pp. 1–48, 1980.

[TEO 11] TEO M., "The liquidity risk of liquid hedge funds", *Journal of Financial Economics*, vol. 100, pp. 24–44, 2011.

[UHL 05] UHLIG H., "What are the effects of monetary policy on output? Results from an agnostic identification procedure", *Journal of Monetary Economics*, vol. 52, pp. 381–419, 2005.

Index

L, M, N

P, R

S

T, V, W

Printed in the United States
By Bookmasters